Advances in Industrial Control

For further volumes:
http://www.springer.com/series/1412

João M. Lemos · Rui Neves-Silva
José M. Igreja

Adaptive Control of Solar Energy Collector Systems

Springer

João M. Lemos
INESC-ID IST
University of Lisbon
Lisboa
Portugal

José M. Igreja
Institute of Engineering of Lisbon
Lisboa
Portugal

Rui Neves-Silva
FCT
New University of Lisbon
Caparica
Portugal

ISSN 1430-9491 ISSN 2193-1577 (electronic)
ISBN 978-3-319-06852-7 ISBN 978-3-319-06853-4 (eBook)
DOI 10.1007/978-3-319-06853-4
Springer Cham Heidelberg New York Dordrecht London

Library of Congress Control Number: 2014940059

Printed on acid-free paper

Springer is part of Springer Science+Business Media (www.springer.com)

This book is dedicated by the authors to Edoardo Mosca, for the scientific influence he had on them,
Eduardo Camacho, for introducing them to solar energy control,
Virgílio Cardoso Igreja, a pioneer of control in Portugal

Series Editors' Foreword

The series *Advances in Industrial Control* aims to report and encourage technology transfer in control engineering. The rapid development of control technology has an impact on all areas of the control discipline. New theory, new controllers, actuators, sensors, new industrial processes, computer methods, new applications, new philosophies…, new challenges. Much of this development work resides in industrial reports, feasibility study papers and the reports of advanced collaborative projects. The series offers an opportunity for researchers to present an extended exposition of such new work in all aspects of industrial control for wider and rapid dissemination.

Solar energy collectors are rapidly becoming a common sight in the fabric of urban infrastructure. Many domestic homes are now graced with either water-based heat-transfer panels or photovoltaic solar panels on their rooftops. In parallel with these myriad small-scale developments, there are proposals and installations already active for the industrial-scale harvesting of solar energy. Indeed, to ensure a significant contribution to the renewable energy mix and to enable the generation of useful amounts of electrical power, industrial-scale plants are essential.

There are two main process architectures for large-scale *thermal* solar-energy-capture systems; either locally focused radiant solar energy is transferred to a heat-transfer fluid as in the "parabolic trough" process or the solar radiant energy is globally focused onto a central heat transfer mechanism as in a central (tower) receiver system. It is the parabolic trough process used in a distributed-collector solar field (DCSF) plant that is of interest in this monograph *Control of Solar Energy Collector Systems* by authors João M. Lemos, Rui Neves-Silva and José M. Igreja.

Professor Lemos and his colleagues open their monograph with an introductory chapter that briefly reviews thermal solar plants before focusing on DCSF plants and their control. This opening chapter also includes a brief review of the control literature for these plants and a discussion of the organisation of the monograph, both of which are very helpful to the reader.

A key feature of this type of plant is that it is a distributed-parameter system with spatially dependent variables, subject to model uncertainty and various types of external disturbances. The monograph includes a useful discussion of a number of processes from industry and elsewhere that belong to this class of systems. These include moisture control, traffic-highway control, glass-tube manufacture, heating fans, steam-super-heater control and arc welding. Thus, the authors' methods and approaches to the control of distributed dynamical systems should be of interest to a wider readership than one simply interested in solar energy collector control.

This monograph is a valuable contribution to the literature on solar-energy-collector-plant control. The theory and experimental results presented also demonstrate the value of the MUltiStep Multivariable Adaptive Regulator (MUSMAR) algorithm that was developed by Prof. Lemos. The outcomes of control approaches that assume either a black-box model or alternatively use physically based process models are also illustrated. Finally, the important links with other industrial processes and fields make this a significant entry to the *Advances in Industrial Control* series.

For the reader interested in the wider solar energy field, the *Advances in Industrial Control* series is very fortunate to have the more encyclopaedic text, *Control of Solar Energy Systems* by Eduardo F. Camacho, Manuel Berenguel, Francisco R. Rubio and Diego Martínez (ISBN 978-0-85729-915-4, 2012) within its imprint. The control of renewable-energy processes is a growing field and the Series Editors are always pleased to receive new proposals for monograph contributions to this ever-growing field.

Glasgow, Scotland, UK

M. J. Grimble
M. A. Johnson

Foreword

The use of renewable energy, such as solar energy, experienced a great impulse during the second half of the 1970s just after the first big oil crisis. At that time, economic issues were the most important factor and so interest in these types of processes decreased when oil prices fell. There is renewed interest in the use of renewable energies nowadays, driven by the need for reducing the high environmental impact produced by the use of fossil energy systems.

One of the greatest scientific and technological opportunities facing us today is to develop efficient ways to collect, convert, store solar energy fundamentals and utilise solar energy at affordable costs. However, there are two main drawbacks of solar energy systems: (i) the resulting energy costs are not competitive yet and (ii) solar energy is not always available when needed. Considerable research efforts are being devoted to techniques which may help to overcome these drawbacks; control is one of these techniques.

A thermal solar power plant basically consists of a system where the solar energy is collected, then concentrated and finally transferred to a fluid. The thermal energy of the hot fluid is then used for different purposes such as generating electricity, sea water desalination, etc. While in other power generating processes, the main source of energy (the fuel) can be manipulated as it is used as the main control variable, in solar energy systems, the main source of power (solar radiation) cannot be manipulated. Furthermore, it changes in a seasonal and on a daily basis, acting as a disturbance when considering it from a control point of view.

These types of plants have all the characteristics needed for using advanced control strategies able to cope with changing dynamics, (nonlinearities and uncertainties). In order to provide viable power production, parabolic troughs have to achieve their task despite fluctuations in energy input, i.e. solar radiation. An effective control scheme is needed to provide operating requirements of a solar power plant.

During the last 30 years considerable effort has been devoted by many researchers to improve the efficiency of solar thermal power plants with distributed collectors from the control and optimization viewpoints. The group of Prof. João M. Lemos has made a very significant amount of this effort.

The main idea behind adaptive control is to modify the controller when process dynamics changes. It can be said that adaptive control is a special kind of nonlinear control where the state variables can be separated into two groups moving in two different timescales. The state variables which change more rapidly, correspond to the process variables (internal loop), while the state components which change more slowly, correspond to the estimated process (or controller) parameters.

Since Åström and Wittenmark first introduced self-tuning control, many works have appeared in the literature describing new methodological developments and applications. Although some of these applications come from industry and there are some commercial adaptive controllers on the market, the technique has not made it to industry as expected, and even many controllers with adaptive capabilities are working as fixed controllers. The main reason for this slow progression of adaptive control in industry is that adaptive control, as other advanced control techniques, requires more knowledge from process operators as it has more tuning parameters and more concepts to be understood. Another fundamental problem is that when applied to a real process the two timescales are not as far apart as they should be and this may be a problem for parameter identification and for assessing stability. It was shown at the beginning of the 1980s that adaptive control systems could be unstabilized by small external perturbations or small modelling errors. Robustness studies of adaptive control systems are, therefore, of paramount importance.

This book presents techniques to model and control solar collector systems. The reader will find a good introduction to different types of industrial solar thermal plants. The modelling of those plants is tackled rigorously and, in particular, the distributed solar collector plants modeling. This book shows developed and probed adaptive model predictive controllers (GPC, MUSMAR), not only through simulation but also through real experiments. Both fixed and variable operation points are considered with several adaptive controllers designed with a bank of linear models.

First, the adaptive control is tackled by linear models and afterwards by nonlinear models that are calculated directly or using feedback linearization. The book ends with a chapter dedicated to tracking variables references. Experimental results on distributed collector solar fields as well as on plants with similar dynamics are described, providing practical examples that serve as a guideline to configure the controllers.

The systematic approach presented in the book will provide a helpful framework for the reader and constitutes an invaluable material for researchers and/or users of solar systems. The book is mainly aimed at practitioners, both from the solar energy community and the control engineering community, although it can be followed by a wide range of readers.

Sevilla, February 2014 Francisco Rubio

Preface

This book addresses methods for adaptive control of distributed collector solar fields (DCSFs). These systems are plants that collect solar energy and deliver it in thermal form. Essentially, they are made of a pipe located at the focus of multiple rows of parabolic concentrating mirrors. The shape of this structure led to the name "parabolic trough" to designate the technology. Inside the pipe circulates a fluid (high temperature resistant oil or molten salt) that is heated by solar radiation, thereby accumulating energy. The thermal energy stored in the fluid may then be used for several purposes, such as water desalination or electric power generation in a boiler/turbine plant.

Since solar thermal plants are a "clean" source of renewable energy, the interest on DCSFs is growing steadily for obvious reasons. Indeed, in recent years, a significant number of commercial solar thermal plants of the so-called "parabolic trough" has been built throughout the world.

In addition to this practical interest, there is a scientific motivation stemming from the following issues:

1. They are distributed parameter systems, i.e. their dynamics depends on space as well as on time;
2. Their dynamics is nonlinear, with a bilinear structure;
3. They are uncertain systems, albeit with a structure which can be explored to design adaptive controllers that yield an improved performance with respect to "black box" approaches to control.

This book aims to present methods for controller design that meet the difficulties associated with these features, being strongly motivated by the specific dynamical structure of DCSFs. Considering the degree of uncertainty in plant dynamics knowledge, we focus on adaptive methods. Furthermore, different design methods are described that assume various degrees of knowledge about the plant structure. As such, some chapters are devoted to design methods that assume only a very limited knowledge about the plant. In order to improve performance, other chapters address methods that rely on the knowledge of a model, simple enough to be useful for controller design, but that captures the dominant plant dynamical features.

Although the book is about control of DCSFs, it does not fully address its automation in detail, or the optimization of its efficiency from a constructive point of view. Instead, the book concentrates on the problem that consists of, given a DCSF, how it can be operated so as to perform the correct manoeuvres, ensuring safe operating conditions and maximizing efficiency. The point of view assumed is the one of the new emerging area of cyber-physical systems, in which mathematical algorithms interact with physical systems in order to achieve a goal, in this case energy production.

The book provides a number of examples of plants that bear resemblances with DCSFs from a structural point of view when their mathematical models are considered. These examples help to explain how the algorithms described can be used in other contexts and also to illustrate their limitations and tuning. Indeed, an interesting feature consists of the fact that the approaches considered can be applied to other plants described by hyperbolic partial differential equations, in particular process plants in which transport phenomena occur, such as dryers, steam super-heaters or even, among other examples, traffic in highways.

Obviously, an important group of readers for this book are control designers interested in DCSFs. Furthermore, by the reasons explained above, this book can also be a source of inspiration for engineers dealing with such processes.

From an "academic" point of view, a nice feature of distributed solar fields is that their dynamics is rich enough to put challenges to the control designer, but at the same time simple enough to allow analytic work to be done in order to study the dynamics and use the conclusions reached to design nonlinear controllers. As such, the book is expected to be of considerable interest to researchers in control and to postgraduate and even advanced undergraduate students of Control.

Some parts of the book, especially the chapter devoted to dynamics and the design methods that directly rely on system nonlinear models may be of interest to applied mathematicians since they provide a case study and a source problem to illustrate other methods.

An important example, repeatedly used throughout the text concerns an experimental distributed collector solar field of Plataforma Solar de Almeria, located in the South of Spain. Indeed, a significant feature of the book consists of the fact that the control algorithms described are illustrated with actual experimental results performed in this plant.

The material covered is supported by several research projects, developed over many years, whose results have been published in peer reviewed journals and conference proceedings and in Ph. D. theses. As such, a number of people in addition to the authors contributed to the experimental results and the corresponding underlying ideas. The main contributions in this respect were made by F. V. Coito, who was involved in the initial experiments with MUSMAR, L. M. Rato, who contributed to dynamic weights in MUSMAR and the application of Switched Multiple Model Adaptive Control, L. Marreiros who cooperated in a case study with an air heating fan and M. Barão who contributed to feedback linearization and the use of control Lyapunov functions. The data concerning solar furnaces were obtained in cooperation with B. A. Costa.

INESC-ID provided the first author (J. M. L.) with the conditions, and the means, to work on this book. The book was partly written within the scope of the projects SFERA II (project number 312643) and STAGE-STE (project number 609837) supported by the European Commission under the FP7 programme. Part of the work was supported by contract PEst-OE/EEI/LA0021/2013.

Finally, the authors must acknowledge and thank the momentum received to complete this work provided by W. A. Mozart, L. van Beethoven, F. Chopin, F. Liszt, N. Paganini, F. Schubert, R. Schumann, F. Mendelssohn Bartholdy and P. I. Tchaikovsky. Without their multidimensional stimulation and encouragement this book could have never been finished.

Lisbon, Portugal, 2014

João M. Lemos
Rui Neves-Silva
José M. Igreja

Contents

Acronyms and Nomenclature

Acronyms

ARMAX	Auto-regressive, moving average with exogenous input (model)
ARX	Auto-regressive with exogenous input (model)
CEP	Certainty equivalence principle
CNRS	Centre National de la Recherche Scientifique
DCSF	Distributed collector solar field
DSG	Direct steam generation
GPC	Generalized predictive controller
LMI	Linear matrix inequality
LPV	Linear parameter varying
LQ	Linear quadratic (controller)
LQG	Linear quadratic gaussian (controller)
MMAC	Multiple model adaptive controller
MPC	Model predictive controller
MUSMAR	Multistep, multivariable, adaptive regulator
MV	Manipulated variable
NMPC	Nonlinear model predictive controller
OCM	Orthogonal collocation method
ODE	Ordinary differential equation
PDE	Partial differential equation
PI	Proportional-derivative (controller)
PID	Proportional-integral-derivative (controller)
PSA	Plataforma Solar de Almeria
RHC	Receding horizon control
RLS	Recursive Least Squares
SIORHC	Stable input/output receding horizon controller

Nomenclature

Model Variables

t	Time, [s]
z	Space dimension measured along the pipe, [m]
T	Fluid temperature inside the pipe, [°C]
Δz, δ	Small increment in z, [m]
Δt, Δ	Small increment in t, [s]
ρ_f	Fluid volumic mass, [kg/m^3]
A_f	Pipe cross section area, [m^2]
c_f	Fluid specific heat, [°C/kg]
R	Solar radiation power, [W/m^2]
$\bar{u}$	Fluid flow, [m^3/s]
u	Fluid velocity, [m/s]
$\bar{\alpha}$	Coefficient of radiation absorption efficiency
α	Normalized coefficient of radiation absorption efficiency
T_a	Ambient temperature, [°C]
L	Pipe length, [m]

Chapter 1
Solar Energy Collector Systems

This chapter provides a broad overview of solar thermal energy systems. The aim is to describe the context of distributed collector solar fields used in plants that apply parabolic trough technology. Furthermore, the temperature control problem associated to distributed collector solar fields is explained and the use of adaptive control to solve it is motivated. In contrast to subsequent chapters that use mathematics to explain control algorithms and the models upon which they rely, the approach in this chapter resorts to qualitative explanations together with some plant data to introduce the main ideas.

1.1 Solar Energy

Solar energy is an important form of renewable energy resource, that is, sources of energy that do not depend on exhaustable stocks. As such, it is receiving an increasing research and development interest. In broad terms, solar energy harvesting plants may be divided into two classes, each associated with different technologies (Boyle 2004; Miller et al. 2010):

- Photovoltaic;
- Thermal.

Photovoltaic technology produces electrical power from solar radiation by exploiting the so-called photovoltaic effect present in some semiconductors. Although widely disseminated and raising interesting issues from the control and optimization points of view, this is not the focus of this book and is not considered further here. The interested reader may consult (Krauter 2010; Green 2006; Rekioua and Matagne 2012) for further information.

The solar thermal systems of concern in this book transform solar radiation into heat that is accumulated in a fluid. Using a heat exchanger, this thermal energy may then be used for different purposes such as water desalination, steam production

J. M. Lemos et al., *Adaptive Control of Solar Energy Collector Systems*, Advances in Industrial Control, DOI: 10.1007/978-3-319-06853-4_1,

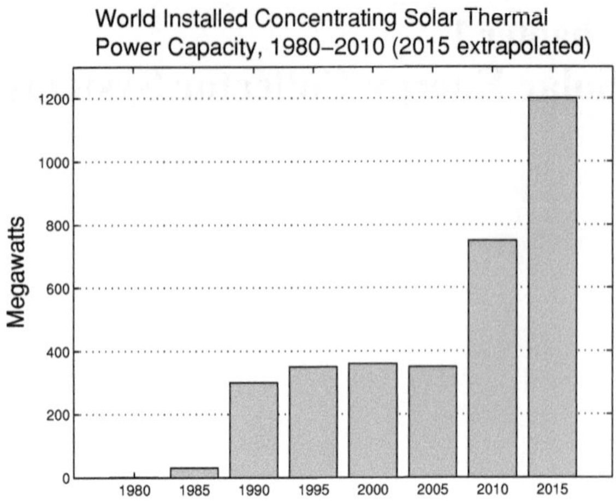

Fig. 1.1 World installed concentrating solar thermal power capacity (*Source* Data highlights on solar energy, March 12, 2010, Earth Policy Institute)

for turbine propelling or even applications such as milk pasteurization and air conditioning.

Needless to say, the interest in solar energy harvesting systems increases during periods of economic crisis and is boosted by factors such as high prices of oil as well as other hydrocarbon fuels or growing environment concerns. In 2005, solar thermal energy accounted for 6.83 % of total world renewable energy and photovoltaic for 0.42 % and these figures are increasing.

Driven by these forces, many national or local administrations are currently promoting the use of solar energy at different levels, from home to the industrial scale. Correspondingly, a number of companies are advertising products for harnessing solar energy, in particular to produce thermal energy. Even large companies that traditionally sell automation and control equipment for "classic" thermoelectric power plants are now offering in their range of products lines devoted to solar thermal energy or to thermal power plants with solar energy contributions.

Figure 1.1 shows the world installed concentrating solar thermal power capacity. Although progress stagnated from 1990 till 2005, many new projects have been started since then. At the industrial level, a number of major projects have been developed and are currently in progress. In March 2009, the 100 MW Andasol plant located in Sierra Nevada in southern Spain started operation, with an installed capacity of 100 MWe. This was the first solar plant in Europe to use parabolic trough technology (explained below), and the second in the world, after the 64 MWe Nevada Solar One in the United States. A thermal storage system uses a molten salt mixture to store solar energy captured during the day and allows the operating time to double.

Another example of a large solar energy generating facility using parabolic trough technology is the Solar Energy Generating Systems (SEGS), consisting of a network of nine units with a total capacity of 354 MWe, located in the Mojave desert, in the United States.

Many other examples of plants starting operation in 2009/2010 or under construction could be indicated, with Spain and the United States leading the list of installed capacity. Most of them use parabolic trough technology (to which this book is devoted) or, to a less extent, solar power towers.

1.2 Types of Solar Thermal Plants

Solar collectors may be classified according to temperature in low, medium, or high temperature. Low and medium temperature collectors are made of flat plates, with a black background and covered with glass to create greenhouse effect. The fluid to be heated passes through a pipe that is serpentine inside the plate. Their applications concern mainly home uses such as swimming-pool heating, for low temperature, or building heating or air conditioning, for medium temperature.

High temperature collectors concern instead industrial uses, such as energy production or material treatment. They include (Miller et al. 2010):

- Distributed collector solar fields;
- Direct Steam generation
- Solar power towers
- Solar furnaces

To this list, one could also add parabolic concentration dishes coupled to Stirling or steam engines. These are omitted since the description of their operation is outside the scope of the methods considered in this book.

Distributed collector solar fields are the focus of this book. From a practical point of view, they are of paramount importance since, as mentioned above, many plants already in operation, or under construction use the parabolic trough technology. Some of the control methods described in this book may also be applied to the other systems. Hence, hereafter we present a concise description of the principles of operation of the different types of thermal systems, with an emphasis on distributed collector solar fields.

1.2.1 Distributed Collector Fields

Distributed collector fields form the core of plants using parabolic trough technology. They are made of parabolic mirrors that concentrate direct incident sun light onto a pipe where a fluid flows. Figure 1.3 shows an example of a dismounted concentration mirror. Another alternative is to use Fresnel reflectors that allows reduced construction costs by the possibility it offers to share a common pipe for several reflectors. Figure 1.2 shows a schematic view of two reflectors (labeled R_1 and R_2) that share the same focal line.

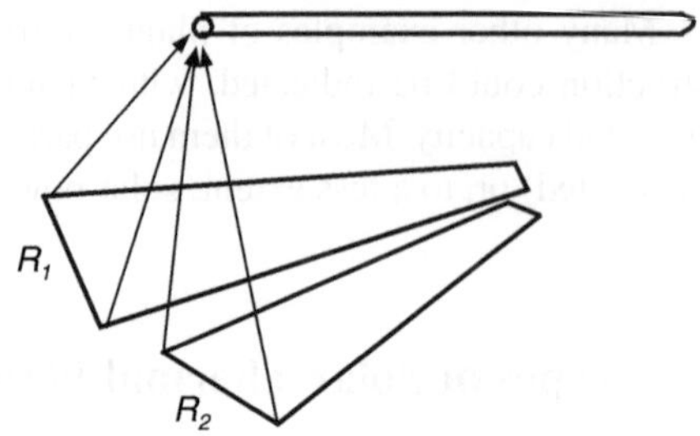

Fig. 1.2 A schematic view of two Fresnel concentrating mirrors sharing the same focal line

Fig. 1.3 A dismounted parabolic mirror concentrating the incident light at its focus

The fluid used to accumulate energy in thermal form may be either an oil that can withstand temperatures up to 300 °C or 400 °C, or a molten salt mixture of 60 % sodium nitrate and 40 % potassium nitrate that is able to reach higher temperatures. Besides cost reduction, the use of molten salt has a number of technical benefits and is being used in a number of recent projects. Many plants, however, use oil.

Figure 1.4 shows a distributed collector solar field. The parabolic mirrors are organized in rows oriented along the East–West direction. The elevation of each row of mirrors is adjusted by a sun tracking controller. The sun elevation sensors used for this controller can be seen in Fig. 1.5.

The distributed collector solar field of Plataforma Solar de Almeria (PSA) used to conduct the experiments described in this book,located in the Tabernas desert in the south of Spain is a pilot plant that is used in the examples described in this book. It consists of 480 parabolic mirrors arranged in 20 rows. Each pair of rows forms a loop. The active part of the pipe corresponds to the segment that is heated by the concentrators. The part of the pipe that connects the loops with the storage tank does not contribute to the heating of the plant but is important from a control point of view because, depending on the process variable being controlled, it may include a delay in plant dynamics. This field was build for experimental purposes and so its nominal

Fig. 1.4 A distributed collector solar field

Fig. 1.5 Sun elevation sensors in distributed collector solar fields

maximum power is small, just 0.5 MW. As mentioned before, other fields already in operation in other locations reach 50 MWe or even more when grouped.

As seen in the sketch of Fig. 1.6, the fluid to be heated is extracted at low temperature from the bottom of the storage tank (seen in the background of Fig. 1.4). It passes through the field where it is heated by solar radiation and returns to the tank, where it is injected at the top. In the case of the DCSF of PSA in which the experiments described in this book were conducted, the fluid is an oil that can withstand up to 300 °C. Since the oil has a low heat conductivity, this fluid forms layers of different temperatures (according to the temperature profile of the incoming oil) that

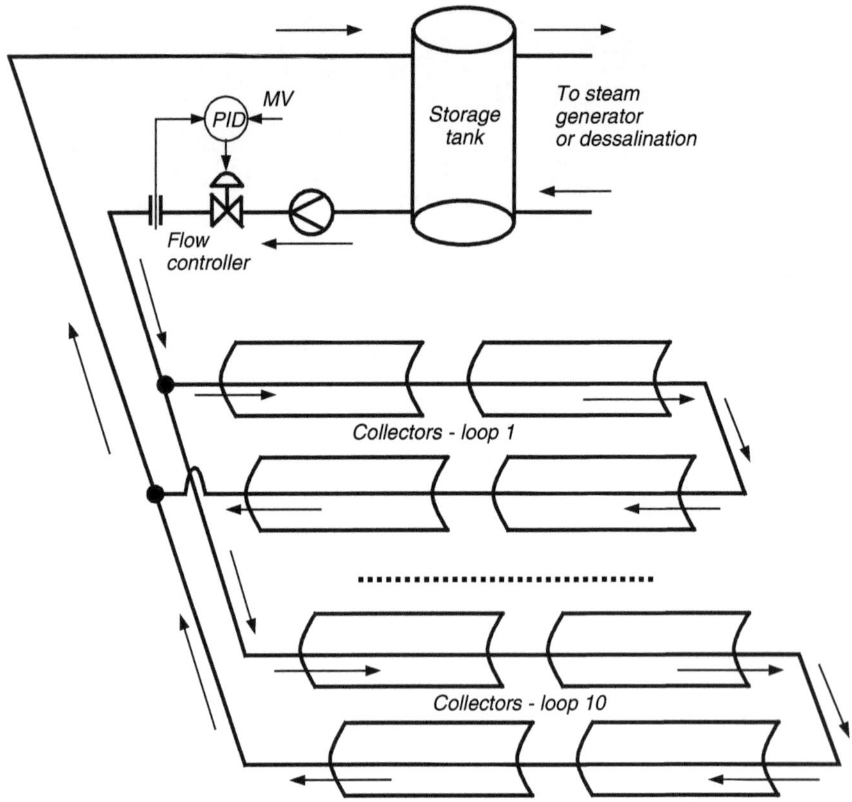

Fig. 1.6 Schematic view of a distributed collector solar field showing the heating fluid circuit

do not mix inside the storage tank. This makes energy storage possible in a single tank where high and low temperature regions coexist.

The flow is ensured by a pump. At PSA, the pump delivers a maximum of 9 l/s. For safety reasons, the minimum flow is imposed to be 2 l/s. In order to manipulate the flow of the thermal working fluid, the outlet pipe from the storage tank is provided with a flow controller, also shown in Fig. 1.6. The flow measure is transmitted to a PID controller that computes the degree of opening of the valve that manipulates flow.

From the top of the storage tank (the hotter zone inside the tank), the oil may then be extracted for use in a desalination plant or to generate steam to feed an electric turbo-generator. Once the energy is extracted, the cool fluid is re-injected at the bottom of the storage tank.

The design of Fig. 1.6 has the advantage of decoupling the collection of solar energy from its use for electricity production. Since energy is stored in thermal form inside the storage tank, the design mentioned allows, for instance, the production of electric power during the night of when solar flux is temporarily interrupted

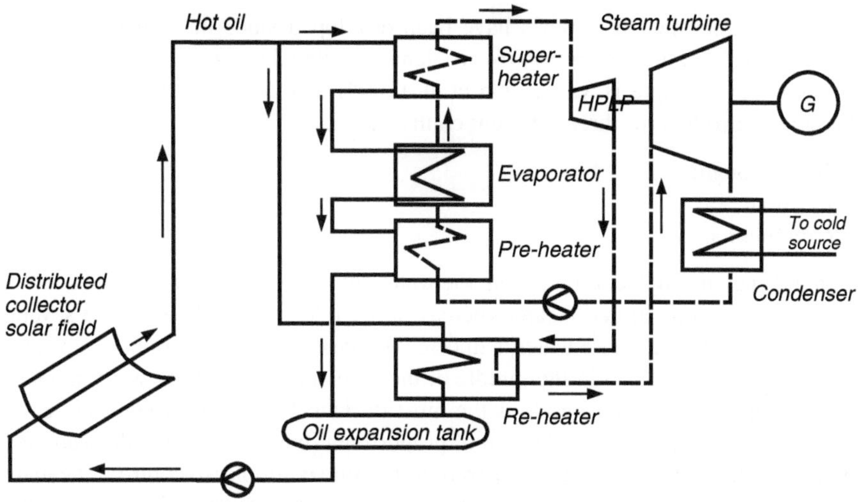

Fig. 1.7 Simplified schematic view of a distributed collector solar field coupled to a turbo-generator

for persistent clouds. A steam generator may however be directly connected to a distributed collector solar field, as in the diagram of Fig. 1.7. In this design, there are two circuits: the fluid circuit and the water/steam circuit. The fluid is heated in the distributed collector solar field and delivers its energy to a steam generator comprising a pre-heater, an evaporator (where the water changes phase to steam) and a super-heater. The steam is then used in a steam turbine coupled to an electric generator. Typical characteristics of the steam delivered to the turbine are a pressure of 104 bar and a temperature of 370 °C. The modeling of the solar field is dealt in this book from a control design point of view (see also the section on literature review, below in this chapter). Modeling of the steam generation part relies on techniques known for thermoelectric power plants, see for example (Ordys et al. 1994; Soares et al. 1997).

1.2.2 Direct Steam Generation

An interesting variant of parabolic trough technology is Direct Steam Generation (DSG). In DSG, water is circulated in the pipe at the collector focus and is directly transformed into steam at the collectors. This results in a simpler overall plant design, requiring lower investment and operation and maintenance costs, and has the advantage of not needing oil, thereby sparing the cost of this element and reducing environmental risks. Furthermore, a higher steam temperature is achieved, together with a higher process efficiency.

Several experimental projects have been developed since the end of 1980s to recently. The literature on the control problems associated to direct steam generation is scarce. An example is Valenzuela et al. (2004) that considers arrangements of classical controllers for three different configurations:

- Once-through;
- Injection;
- Recirculation.

In the once-through configuration, the water entering the receiver tube is directly converted into superheated steam whereas in the injection mode there are several points along the tube where water is injected. In the recirculation structure, a water-steam separator is placed in the middle of the tube and the resulting water is recirculated. The once-through mode is the most difficult to control (Valenzuela et al. 2004, 2005) requiring a combination of proportional-integral (PI), feedforward and cascade control. The recirculation operation structure is simpler to control. The main objective of obtaining steam at a specified constant temperature and pressure at the receiver tube outlet can be achieved by sacrificing the amount of steam produced (Valenzuela et al. 2006).

A number of difficulties require further progress. These include poor control performance in the presence of strong and fast solar radiation transients due to clouds, instabilities of the two-phase flow inside the receiver tubes and the occurrence of major temperature gradients at the receiver pipes.

1.2.3 Solar Towers

Solar tower fields comprise a single tower on top of which a receiver is located. The field surrounding the tower is filled with flat mirrors, called heliostats, that reflect the sun and focus it on the receiver. This is in contrast to distributed collector solar fields, where the mirrors concentrate the sun radiation along a lengthy focus. As a result, the temperature achieved with solar towers is higher with respect to distributed collectors and an increase in the efficiency of the conversion between thermal and electric energy results. A disadvantage consists in the need of adjusting the heliostat positions along two axis, whereas distributed collector fields require only one axis, corresponding to sun elevation. This complicates the design and implies higher construction as well as maintenance costs.

The receiver contains molten salt that is heated to temperatures above 500 °C. Figure 1.8 shows a simplified schematic view of a solar tower plant. There are two main circuits, corresponding to molten salt and water/steam. The molten salt is pumped from the cold molten salt reservoir up to the receiver, where it is heated and then goes to the hot molten salt reservoir. The hot molten salt reservoir acts as an energy storage system. From it, hot molten salt is pumped to provide heat to the steam generator, and then back to the cold reservoir.

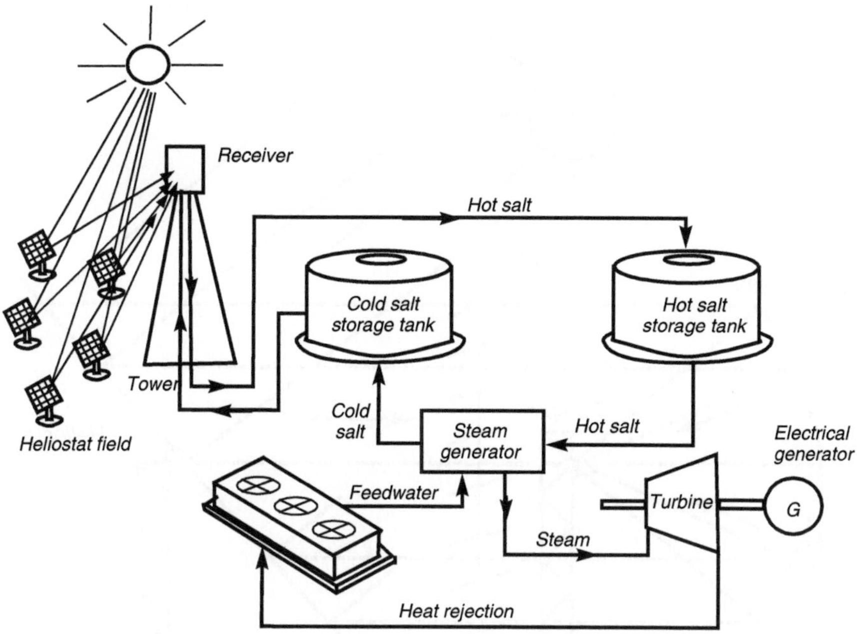

Fig. 1.8 Schematic view a solar tower plant

The water/steam circuit is similar to other plants. The steam produced by the steam generator, which by itself is a subsystem with several parts, is used by the turbine, that can have several bodies that operate at different pressures, with a reheater in-between, to convert thermal energy to mechanical energy that drives the generator.

Although not so common as parabolic trough installations, there are several significant examples of plants of solar power type, as well as companies devoted to their design and construction.

1.2.4 Solar Furnaces

Solar furnaces are highly concentrating plants that yield a high temperature over a limited area, close to their focus. They are mainly used for material testing and they may range from relatively small structures to a very large size, such as the Odeillo furnace of French CNRS plant, located in the Pyrenees.

As shown in the schematic view of Fig. 1.9, a solar furnace is made of the following parts:

- Heliostats
- Shutter

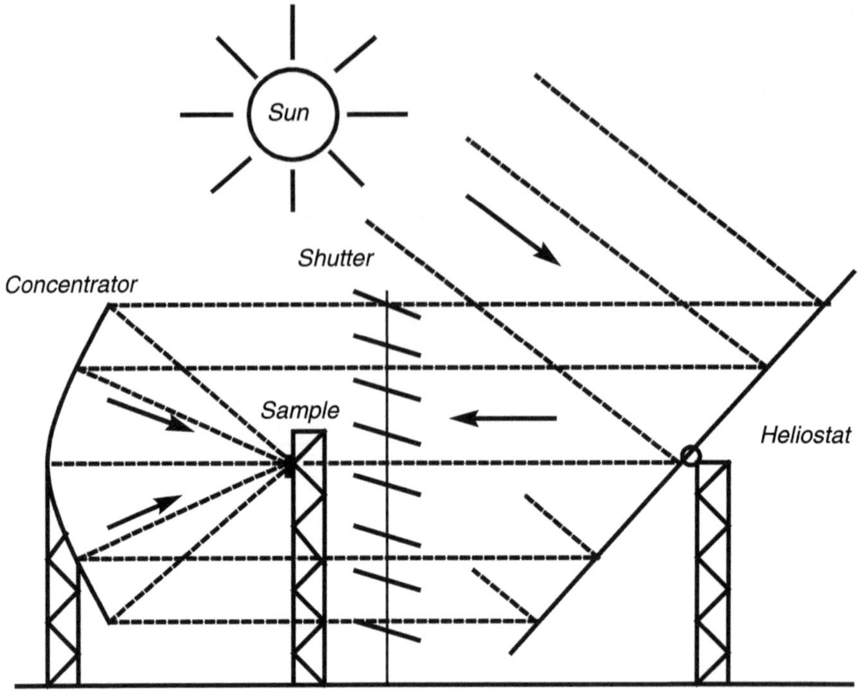

Fig. 1.9 Schematic view of a solar furnace

- Parabolic concentrator
- Sample support

The heliostats are computer-driven flat mirrors that track the sun throughout the day, such that the parabolic concentrator receives the sun beam always along the same direction. The movement of the heliostats may be defined in either one of two modes:

- Open loop control, where the heliostat angles are preprogrammed depending on solar time and the geographic coordinates of the place;
- Closed loop control, in which the angles are selected by a feedback controller that seeks to keep the mirrors oriented in such a way as to maximize the incoming solar radiation flux to the parabolic concentrator.

The solar radiation flux reflected by the heliostats crosses the shutter and reaches the concentrator that focuses it at the area where the sample to be treated is located.

The shutter is the actuator. It is made of blades that rotate along their longitudinal axis, driven by a motor that receives commands from a computer. In some cases, the shutter is located such that it interrupts the radiation flux after the concentrator (but at a point where the temperature is still relatively low). Figure 1.10 shows an example of this kind of shutter. In any case, the shutter is a nonlinear actuator since,

Fig. 1.10 The shutter (the *white* flaps at the center of the image) and the focus of a 6 kW solar furnace where a sample is located handled by a support. The radiation enters from below and the concentrator is located above the shutter

when the shutter aperture command varies from 0 % (fully closed) to 100 % (fully opened) the amount of radiation that passes per unit time is a nonlinear function of the aperture setting. This nonlinearity is exactly known from the shutter geometry and can however be compensated.

The concentrator can be built either of a single parabolic mirror or of multiple mirrors, each one a section of a parabola.

In general terms, the control of the furnace consists of manipulating the shutter such that the temperature at the focus is at the desired value, either kept constant or closely following a specified reference profile.

The dominant dynamics of the furnace results from an energy balance made at the material sample placed at the focus for heating and may be represented by the ordinary differential equation (Berenguel et al. 1999):

$$\dot{T}_s = -\alpha_1(T_s^4 - T_e^4) - \alpha_2(T_s - T_e) + \alpha_3 R \tag{1.1}$$

Here, T_s denotes the temperature of the sample, T_e the temperature of the environment (all expressed in K), R is the incoming solar radiation (expressed in Wm^{-2}) and α_1, α_2 and α_3 are parameters.

The derivative of the sample temperature depends on three terms. The first term, $-\alpha_1(T_s^4 - T_e^4)$, models losses by radiation and corresponds to what is known in physics as Boltzman's Law. Due to the dependence on the fourth power of the temperature, this term embeds a strong nonlinearity in the dynamics, a fact that shapes the type of control used in this type of plant.

The second term, that depends linearly on the temperature, accounts for losses due to conduction and convection. Finally, the third term represents the incoming energy from solar radiation. This is modulated by the shutter position. Since the shutter is a moving mass acted by an electric motor, it also contributes to the dynamic behavior of the system. However, there is a time scale separation between the electromechanical part (the shutter and its electromechanical driver, which is faster) and the thermal part (which is slower) that can be taken advantage of for controller design.

When comparing the dynamics of solar furnaces and distributed collector solar fields, some differences are apparent:

- While the dynamics of solar furnaces are concentrated (that is, described by ordinary differential equations) and low order (mainly a first order system plus the shutter dynamics), the dynamics of distributed collector solar fields depend on spacial variables, as well as on time and are described by partial differential equations;
- Solar furnaces work at much higher temperatures than distributed collector solar fields and are built such that the losses by radiation must be accounted for. As mentioned above, this implies a dependence on the fourth power of temperature which is a strong nonlinearity. As will be explained below, distributed collector solar fields present a different type of nonlinearity, where the state multiplies the manipulated variable.

Although solar furnaces are usually modeled by lumped parameter models, it should be recognized that this is an approximation. The spacial distribution of the incident radiation over the sample being processed is not uniform and one has to specify what is the temperature that is actually being controlled. For instance, even in small samples, a pyrometer measuring the temperature of the upper surface of the sample and a thermocouple measuring the temperature in the lower surface yield different values and with different time profiles. Figure 1.11 shows an example.

Despite its importance and scientific interest, the literature on control of solar furnaces is scarce (Berenguel et al. 1999; Costa and Lemos 2009; Costa et al. 2011; Lacasa et al. 2006; Paradkar et al. 2002). Although it is possible to use simple PID controllers with acceptable results, as shown in the example of Fig. 1.12, there is considerable room for improvement using advanced control techniques.

As in other types of solar plants feedforward from solar radiation greatly improves tracking performance (Berenguel et al. 1999), a fact that is specially important in solar furnaces used for material testing where imposing a specified profile of temperature is quite important. This is no surprise since, as is well known and explored in different types of processes such as chemical (Stephanopoulos 1984) or electromechanical processes, feedforward anticipates the existence of tracking errors due to accessible disturbances and acts to prevent, or at least reduce their appearance. The design of

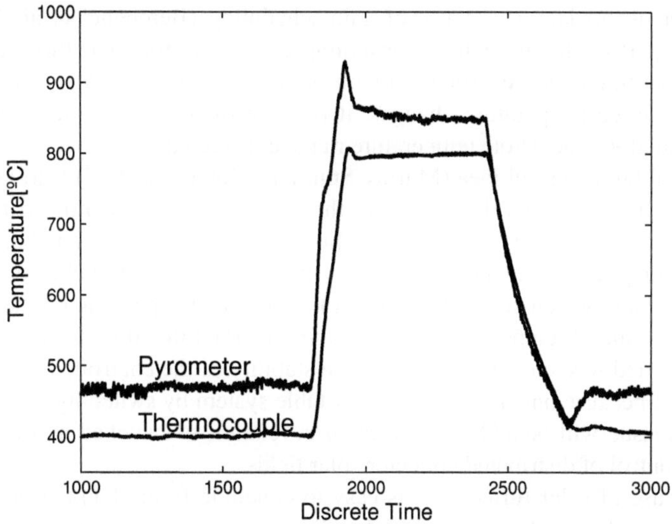

Fig. 1.11 Measurement of the *upper side* of the sample temperature with a pyrometer and of the *lower side* of the sample temperature with a thermocouple in 6 kW solar furnace

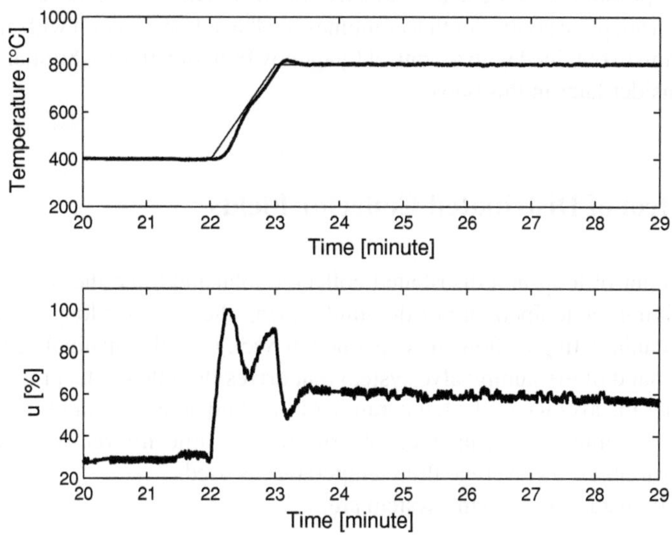

Fig. 1.12 Temperature control of a 6 kW solar furnace

a feedforward controller implies, off course, that a reliable model of the plant is available, including the effect of both the manipulated variable (shutter command) and accessible disturbance (incoming solar radiation) on the output to be controlled (sample temperature).

Another useful idea is the use of gain-scheduling (Berenguel et al. 1999). A significant part of the difficulty of designing controllers for solar furnaces comes from the fact that they are strongly nonlinear. When performing tests on the stress induced by large temperature changes, it is not possible to consider a linearized model valid along the whole temperature range considered.

A natural but powerful idea (Murray-Smith and Johansen 1997) is to use a pencil of local linear models that represent the plant dynamics for limited ranges of temperature. To each local model, a local controller is then associated and a suitable switching or interpolation supervisory mechanism is then included to decide at every moment which controller(s) is(are) to be active depending on the value of the current value of temperature. It should be remarked that this supervisor should be programmed in such a way to ensure the stability of the controlled system. This is nontrivial because one may create an unstable system by switching among stable ones. This issue will be addressed later on in relation to switched multiple model adaptive control of distributed collector solar fields.

In the case of solar furnaces, it is easy to conclude from (1.1) that local models are simply first order linear models whose gain and time constant vary with the equilibrium temperature. Hence, selecting PID's as local controllers allows the corresponding closed loop poles to be placed appropriately.

Another possibility is the use of adaptive control. The Costa and Lemos (2009) describes an adaptive controller that combines feedback linearization with an adaptation mechanism obtained with a Control Lyapunov Function (CLF). Similar methods will be consider later in this book.

1.3 Control of Distributed Collector Fields

The main control loop in a distributed collector solar field like the one of Fig. 1.4 aims at driving the temperature of the fluid leaving the collector loops to a desired value, by manipulating its flow. As explained in Sect. 1.2.1 the manipulated variable is the command of the pump/valve system that drives fluid flow. The process output can be either the average of the temperature of the fluid at the output of the collector loops or its maximum. Since in practical terms there is little difference in controlling these two variables, the average fluid temperature is used.

The main disturbances to the system are:

- Changes in incoming solar radiation due to passing clouds and moisture scattered in the atmosphere;
- Changes in the temperature of the fluid at the input of the pipe;
- Changes in ambient temperature;
- Disturbances due to various factors such as dust deposition on the mirrors and wind affecting the shape of the mirrors.

The first three types of disturbances in the above list can be measured and used as feedforward signals. In particular, feedforward from solar radiation is quite important

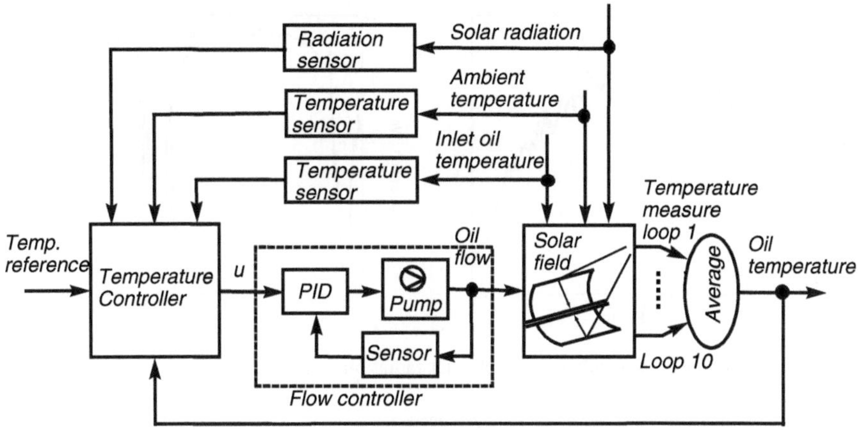

Fig. 1.13 Basic temperature control system for a distributed collector field

to improve the performance. The other two disturbances are less important. It should be remarked that the measure of incoming solar radiation may present difficulties due to the large size of the field that may lead to situations in which the collectors are not uniformly illuminated.

Figure 1.14 shows two examples of daily records of solar radiation. Due to the Sun's apparent movement, the intensity of radiation arriving at a given point on Earth has a sinusoidal evolution with the hour of the day. As is well known, its peak value depends on the period of the year, for details see for example, Tiwari (2002) or any other reference on solar energy. Superimposed on this deterministic variation, there are two main stochastic fading effects. One is associated with moisture scattered in the atmosphere. By absorbing radiation in a random way, it causes fast small amplitude changes of the intensity of the radiation arriving to the collector field. The fluctuations on the record on the left of Fig. 1.14 are predominantly due to moisture. The other factor that reduces radiation intensity are passing clouds. The record on the right of Fig. 1.14 shows three groups of clouds passing that drastically reduce the arriving solar radiation intensity, sometime to values close to just 100 Wm^{-2}.

It should be remarked that, specially at the beginning and at the end of the day, the radiation intensity varies such as to resemble a ramp function. Hence, basic system theory arguments allow us to conclude that if the controller contains just one integrator a steady state error will result. This will be made apparent later in some experimental results.

Given the above definition of manipulated, process and disturbance variables, the block diagram of a distributed collector solar field controller is as in Fig. 1.13. In many cases, feedforward terms from ambient temperature and from inlet fluid temperature are not used. It is remarked that the inlet fluid temperature is affected by the load on the system, that is, by the amount of energy per unit time extracted from the fluid by the secondary circuit connected to the equipment that consumes power.

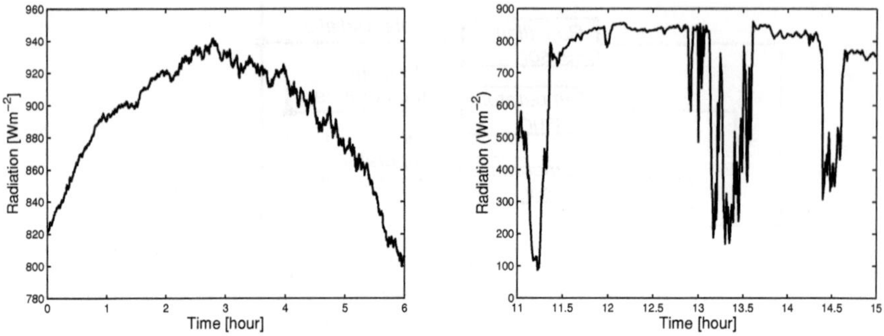

Fig. 1.14 Examples of two daily records of solar radiation

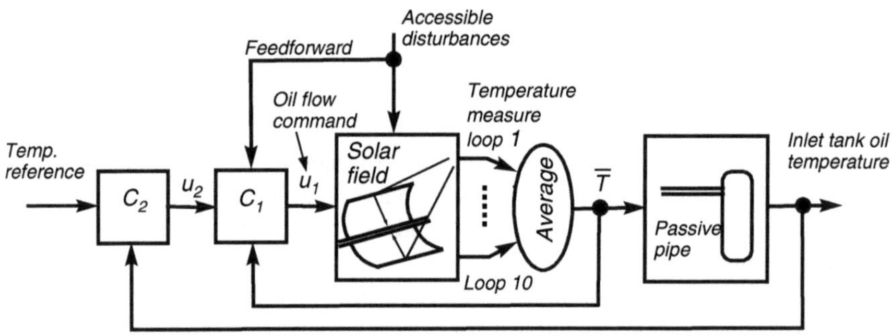

Fig. 1.15 Cascade temperature control system for a distributed collector field

An alternative structure uses two feedback loops connected in cascade as shown in Fig. 1.15 (for the sake of simplicity the block diagram in this figure omits the sensors and the fluid flow control system). The inner loop (associated with controller C_1 in Fig. 1.15) controls the average temperature $\bar{T}$ of the fluid leaving the collector loops and includes the feedforward actions as explained above. Its manipulated variable is the fluid flow command. The outer loop (associated with controller C_2) uses as manipulated variable, the temperature set-point of the inner-loop (denoted u_2 in Fig. 1.15) and as the process variable to be controlled, the fluid temperature at the inlet to the storage tank. In-between the end of the collector loops and the storage tank there is a pipe that collects heated fluid from the collector loops and transports it to the storage tank.

The motivation for using cascade control is common to what is found in other processes (Stephanopoulos 1984). In addition to decoupling the design of controllers for different parts of the plant, it reduces the effect of disturbances affecting the inner loop, thereby decreasing their effect in the outer-loop and improving tracking performance.

The dynamics of the relation between the average temperature at the output of the collector loops and the temperature at the storage tank inlet is mainly a delay that varies with the average value of fluid flow. This means that, if this outer loop is to

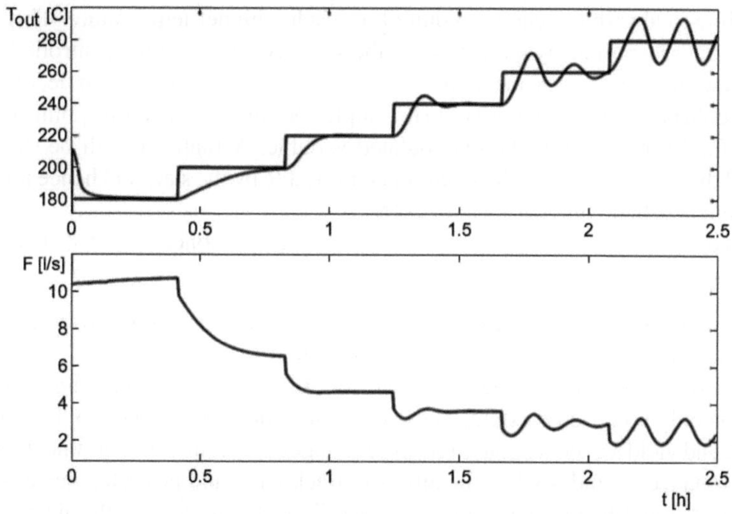

Fig. 1.16 Simulation of a distributed collector field controlled by a PID, demonstrating non-linear behaviour. *Bottom plot* The manipulated variable. *Top plot* Outlet fluid temperature and the corresponding reference

be closed with a controller that is robust with respect to changes in plant delay. A suitable controller is provided by some Model Predictive Control algorithms. More details on adaptive cascade control solutions will be provided in Chap. 3.

In general, unless explicitly stated otherwise, when we consider the control of the field, we refer to the situation of Fig. 1.13 and not to cascade control. Furthermore, we refer to the average of the temperatures of the fluid leaving the collector loops simply as "the temperature".

A detailed discussion of the dominant dynamics of distributed collector solar fields is made in Chap. 2. Here, based on experimental data as well as simulations, we just point out a number of characteristic features of distributed collector solar fields that motivate the use of adaptive control techniques and identify some major difficulties facing controller design.

The first point to note is that the system is nonlinear. The nonlinearity is shown by the simulation results of Fig. 1.16. These records concern the situation in which the temperature is controlled with a fixed gain PID that is well tuned for a working point close to 200 °C. With this controller, the reference is varied between 180 and 280 °C in five steps of 20 °C each. As can be seen, at lower temperatures the response is sluggish, but, as the working point increases, there is more and more oscillation. Around 280 °C there is a constant amplitude oscillation. The linear constant parameter controller was not able to adequately control the system over the whole operating range due to the inherent system nonlinearity.

In qualitative terms, this can be easily understood. At working points corresponding to higher temperatures, the average fluid flow is smaller because the fluid requires more time to absorb the energy required to reach a higher temperature. Therefore, the system transport delay increases and the stability margin of the controlled linear approximation decreases causing the closed-loop response damping to decrease.

It should be noted that the situation is complex because the operating point depends on the fluid flow which is the manipulated variable. Actually, as will be shown in Chap. 2 there is a product of the manipulated variable by the state and hence this type of system is called "bilinear" (Elliot 2009).

This discussion leads us to the following conclusions concerning distributed collector solar fields:

- For a controller to be effective over a wide range of reference values it must take into account plant nonlinearity and compensate it;
- Adaptive controllers relying on linear control laws can compensate nonlinear effects only up to some extent. These controllers are able to adapt to the working point and yield a good response for *small* changes of the reference around it, but the performance degrades when the aim is to track large and fast reference changes. Chapters 5 and 6 consider nonlinear controllers able to cope with this problem. In particular, Chap. 5 will present experimental results on the response to a 50 °C sudden variation of the reference with virtually no overshoot.

Another important issue is uncertainty. In distributed collector solar fields uncertainty comes from several sources. These include parametric uncertainty and unmodeled terms.

Issues affecting parametric uncertainty are manifold: When identifying a model, parameter estimates become known only up to a certain precision. This is common to all modeling processes. More specific to distributed collector solar fields are the processes that lead to parameters change. One is aging: With the passage of time the working fluid used to capture solar energy may change its thermal characteristics. Another is dust deposition/removal, by factors such as wind or rain, over the surface of concentrating mirrors that changes its reflectivity coefficient.

Apparently equal systems may actually present different behavior when subjected to the same stimulus. An example is provided by Fig. 1.17 that shows the temperature records at the outlet of two different collector loops, made at the same time, with the same fluid flow and under the same solar radiation incidence. Nevertheless, the records are not only different, but the maximum is attained in each of them at different time instants.

Unmodeled terms have also other sources. A major example concerns the fact that the plant is infinite dimensional but that controller design usually rely on finite dimensional models. Furthermore, several subsystems, such as energy accumulation in the glass covering the pipe conveying the heating medium, or the dynamics of the flow controller, are usually neglected and contribute to uncertainty in plant model knowledge. Another example is provided by the difference between the fluid flow command and the actual fluid flow (Fig. 1.18). The temperature controller manipulates the fluid flow command but the fluid flow control loop adds a delay of 15 s which is significant (in the computer control experiments described in this book the sampling time is 15 s).

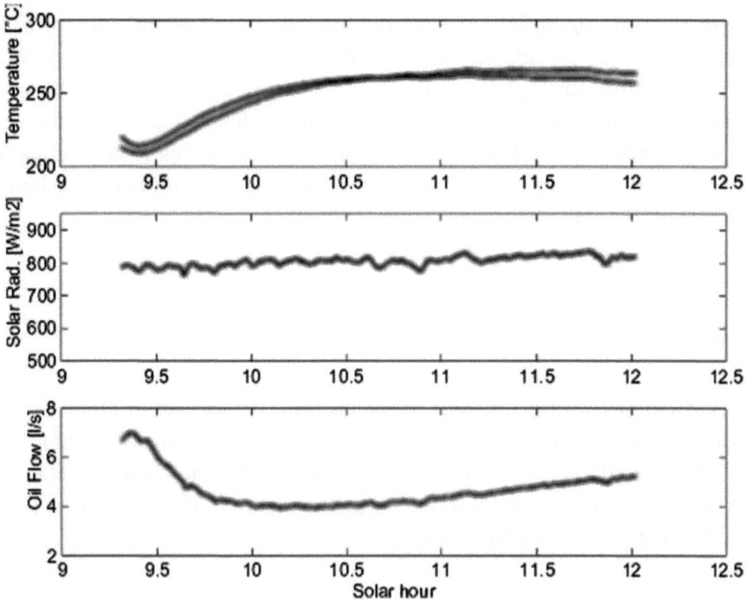

Fig. 1.17 Temperature records at the outlet of two different collector loops in distributed collector field

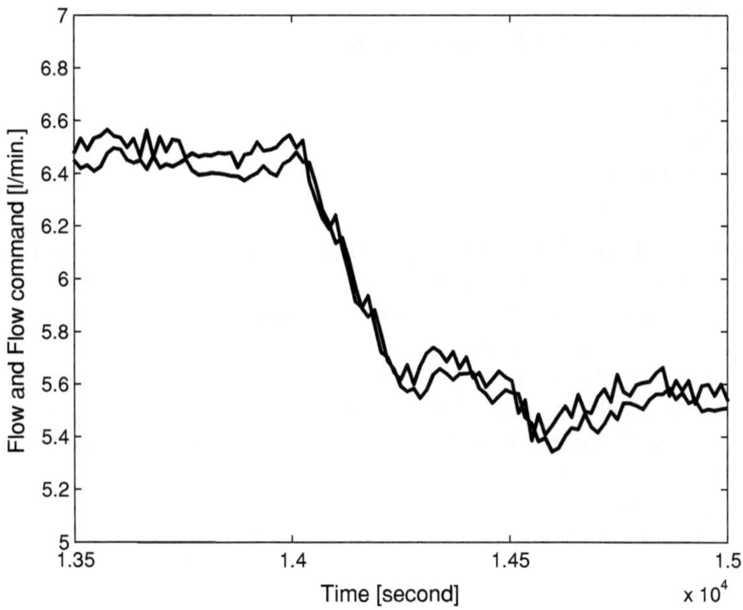

Fig. 1.18 The fluid flow and the command fluid flow signals in an experiment with a distributed collector field

The presence of uncertainty factors motivate the use of adaptive control. Based on current plant data, the adaptation mechanism corrects the controller gains so as to optimize their match to plant dynamics. A word of caution should however be said: Adaptation corresponds to a highly nonlinear feedback that may destabilize the plant if proper care is not taken to produce an adequate design (Ioannou and Fidan 2006). This said, adaptive control algorithms are powerful tools that enhance the performance of distributed collector solar fields.

1.4 Literature Overview

Control of distributed collector solar fields has been the subject of intense research, well documented in many publications. Some overviews are also available such as Lemos (2006) for adaptive control applications to DCSFs and Camacho et al. (1997, 2012, 2007a, b) in a more general framework. The methods described in the literature that have been applied to DCSFs include:

- Optimal Control
- Predictive Control
- Gain scheduling
- Adaptive Control including self-tuning, dual control and Control Lyapunov Functions.
- Switched Multiple Model Adaptive control
- Frequency based control
- Hybrid Control
- Fuzzy Logic Control
- Internal model control
- Nonlinear predictive control

The major role played by changes of solar radiation and plant uncertainty lead to the approach of Camacho et al. (1992) where a pole placement self-tuning controller with a series feed-forward compensator is used. An improvement of the adaptation mechanism and of the underlying control law was possible by resorting to predictive adaptive control techniques. Different forms of adaptive GPC (Camacho et al. 1994; Camacho and Berenguel 1997) and MUSMAR (Coito et al. 1997) were then demonstrated with success. Variants include cascade control (Rato et al. 1997; Silva et al. 1997) and dual control to eliminate start-up transients due to initial adaptation in the absence of good *a priori* knowledge of the plant dynamics (Silva et al. 2005). Another example of adaptive predictive control is described in Pickhardt (2000).

By performing a frequency response analysis under conditions which correspond to linear behavior, it is possible to recognize the occurrence of anti-resonances (Meaburn and Hughes 1993). This is confirmed by a simplified analysis based on the PDE (partial differential equation) describing collector loop dynamics (Berenguel and Camacho 1995) and lead to the design of controllers based on frequency methods (Berenguel and Camacho 1995, 1996; Meaburn and Hughes 1994). In Meaburn and Hughes (1994) a pre-scheduled adaptive controller for resonance cancelation has been presented and in Berenguel and Camacho (1995, 1996) an adaptive control algorithm using an Internal Model Control structure together with a frequency design approach has been introduced. Another internal model based controller is described in Farkas and Vajk (2002), while Normey-Rico et al. (1998) uses adaptive dead-time compensation.

While adaptive control already provides some form of accommodation of nonlinear behavior by adjusting the controller gains according to the operating point, explicit recognition of plant nonlinearities and their exploitation is much likely to improve the ability of the plant to meet control objectives in a wider set of operating conditions. First steps in this direction were made by employing gain scheduled constant parameter GPC (Camacho et al. 1994) as well as other gain scheduled approaches (Johansen et al. 2000) and switched multiple model supervisory controllers (Lemos et al. 2000; Pickardt 2000; Pasamontes et al. 2011).

In Barão et al. (2002), a nonlinear controller is developed which explicitly takes into account the distributed parameter model of the solar plant. A nonlinear adaptive controller was designed using a joint control Lyapunov function for parameter estimation and control. Nonlinear control with feedback linearization is also considered in Igreja et al. (2005) and Cirre et al. (2007). The inclusion of constraints in feedback linearization in relation to distributed collector solar fields is considered in Roca et al. (2009).

Also departing from the partial differential equation model of the plant, Johansen and Storaa (2002) proposes a design based on Lyapunov methods, using internal energy as Lyapunov function. Lyapunov methods, but using a quadratic function, are also the realm of Carotenuto et al. (1985, 1986).

Optimal and suboptimal control laws are considered in Orbach et al. (1981) and Rorres et al. (1980), where an extension of Pontryagin's Maximum Principle to DCSFs is used.

Other approaches include fuzzy logic based control (Rubio et al. 1995; Cardoso et al. 1999; Flores et al. 2005; Pereira and Dourado 2002) and control based on neural network models (Arahal et al. 1998).

A major advance is provided by Silva et al. (2003a, b), in which a time varying sampling interval is used. This approach, to be discussed in detail in this book, implements a change in time scale that linearizes the plant, thereby allowing very high sudden changes in reference with a fast response and no overshoot.

1.5 Book Organization

The book is organized as follows:

This chapter introduces distributed collector solar fields in the context of plants that exploit solar thermal energy. The control problem associated to the plant is formulated and a literature review is made. This chapter also explains in broad terms why adaptive control is important for the application at hand.

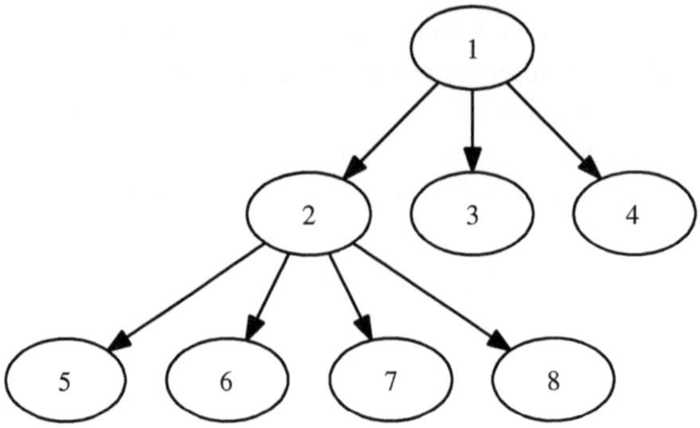

Fig. 1.19 Chapter interdependency

Chapter 2 introduces a reduced complexity model and addresses methods for studying the resulting dynamics. The main result is a discrete time approximation to the evolution over space of the process variable.

The control problem is then considered from a "black box" point of view and insight from the PDE model describing plant dynamics is not used. Instead, controller design is based on linear models which are adapted in order to tackle nonlinear behavior. This is the subject of Chap. 3, concerned with Predictive Adaptive Control, and Chap. 4, where Multiple Model Switching Control is applied.

Although the above approach has advantages in what concerns the speed of developing a controller prototype for application to a new process, it does not take advantage of knowledge about the dominant plant structure. The remaining chapters, therefore, address the problem of how to explore and use this knowledge.

Chapter 5 capitalizes on the results of Chap. 2 to show how to develop a variable sampling interval control algorithm that overcomes the limitations of the black box approach by allowing very fast temperature changes.

Chapter 6 explores an alternative method that uses the field model structure. While the class of algorithms of Chap. 5 relies on a change of the time variable to attain the control objectives, exact feedback linearization and adaptation based on Lyapunov's Second Method are explored here.

Chapter 7 approaches the problem of following a pre-specified reference temperature profile combining flat system concepts for system inversion and adaptive control.

Finally, Chap. 8 draws conclusions and provides a discussion on the interrelation of the adaptive control algorithms presented and their application in the wider context of renewable energy production.

Figure 1.19 shows chapter interdependency. After reading Chap. 1, the reader may take an option. Readers interested in model based design should read Chap. 2, on models and dynamics, and then go to either 5, 6, or 7 that address different forms of model-based adaptive controller design. If, however, the reader's focus of interest is on "black box" methods, the reader may jump directly to either Chap. 3 or 4.

1.6 Main Points of the Chapter

Being a key part of parabolic trough technology for thermal solar plants, distributed collector solar fields (DCSF) are currently receiving much interest for energy production, with many significant projects either in operation or under construction. They are made of parabolic mirrors that concentrate sun light at their focus where a pipe is located. Inside the pipe flows an fluid that is heated, thereby accumulating energy. The main control loop in DCSF regulates the outlet fluid temperature. The manipulated variable is fluid flow and the main disturbances are incoming solar radiation, ambient temperature and inlet fluid temperature. The main control objective consists of adjusting the fluid flow to keep the outlet fluid temperature close to the desired reference. Due to the levels of uncertainty involved, performance may be improved using adaptive control. However, since DCSF are nonlinear plants (actually of bilinear type) the controller design should rely on nonlinear models.

References

Arahal M, Berenguel M, Camacho E (1998) Neural identification applied to predictive control of a solar plant. Control Eng Pract 6:333–344

Barão M, Lemos JM, Silva RN (2002) Reduced complexity adaptive nonlinear control of a distributed collector solar field. J Proc Control 12:131–141

Berenguel E, Camacho EF (1996) Frequency-based adaptive control of systems with antiresonance modes. Control Eng Pract 4(5):677–684

Berenguel M, Camacho EF (1995) Frequency based adaptive control of systems with antiressonance modes. Preprints of 5th IFAC symposium adaptive systems in control and signal processing. Budapest, Hungary, pp 195–200

Berenguel M, Camacho, EF, García-Martín FJ, Rubio FR (1999) Temperature control of a solar furnace. IEEE Control Syst 19(1):8–24

Boyle G (ed) (2004) Renewable energy. Oxford University Press

Camacho EF, Berenguel M (1997) Robust adaptive model predictive control of a solar plant with bounded uncertainties. Int J Adapt Control Signal Proc 11(4):311–325

Camacho EF, Berenguel M, Bordóns C (1994) Adaptive generalized predictive control of a distributed collector field, IEEE Trans Control Syst Technol 2(4):462–467

Camacho EF, Berenguel M, Rubio F (1994) Application of a gain scheduling generalized predictive controller to a solar power plant. Control Eng Pract 2(2):227–238

Camacho EF, Berenguel M, Rubio F (1997) Advanced control of solar plants. Springer

Camacho EF, Berenguel M, Rubio FR, Martínez D (2012) Control of solar energy systems. Springer

Camacho EF, Rubio FR, Berenguel M, Valenzuela L (2007) A survey on control schemes for distributed solar collector fields. Part I: modeling and basic control approaches. Solar Energy 81:1240–1251

Camacho EF, Rubio FR, Berenguel M, Valenzuela L (2007) A survey on control schemes for distributed solar collector fields. Part II: advanced control approaches. Solar Energy 81:1252–1272

Camacho EF, Rubio FR, Hughes FM (1992) Self-tuning control of a solar power plant with a distributed collector field. IEEE Control Syst Mag 72–78

Cardoso A, Henriques J, Dourado A (1999) Fuzzy supervisor and feedforward control of a solar power plant using accessible disturbances. In: Proceedings of European Control Conference, ECC 99
Carotenuto L, La Cava M, Muraca P, Raiconi G (1986) Feedforward control for the distributed parameter model of a solar power plant. Large Scale Syst 11:233–241
Carotenuto L, La Cava M, Raiconi G (1985) Regulator design for the bilinear distributed parameter of a solar power plant. Int J Syst Sci 16:885–900
Cirre CC, Berenguel M, Valenzuela L, Camacho EF (2007) Feedback linearization control for a distributed solar collector field. Control Eng Pract 15:1533–1544
Coito F, Lemos JM, Silva RN, Mosca E (1997) Adaptive control of a solar energy plant: exploiting accessible disturbances. Int J Adapt Control Signal Process (11):327–342
Costa BA, Lemos JM, Rosa LG (2011) Temperature control of a solar furnace for material testing. Int J Syst Sci 42(8):12531264
Costa BA, Lemos JM (2009) An adaptive temperature control for a solar furnace. Control Eng Pract 17:1157–1173
Elliot DL (2009) Bilinear control systems. Springer
Farkas I, Vajk I (2002) Internal model-based controller for a solar plant. In: Proceedings of 15th IFAC World Congress
Flores A, Saez D, Araya J, Berenguel M, Cipriano A (2005) Fuzzy predictive control of a solar power plant. IEEE Trans Fuzzy Syst 13(1):58–68
Green MA (2006) Third generation photovoltaics. Springer
Igreja JM, Lemos JM, Barão M, Silva RN (2005) Adaptive receding horizon control of a distributed collector solar field. In: Proceedings of IEEE CDC-ECC 2005, Sevilla, Spain
Ioannou P, Fidan B (2006) Adaptive control tutorial. SIAM
Johansen TA, Hunt K, Petersen I (2000) Gain-scheduled control of a solar power plant. Control Eng Pract 8(9):1011–1022
Johansen TA, Storaa C (2002) Energy-based control of a solar collector field. Automatica 38:1191–1199
Krauter SCW (2010) Solar electric power generation—photovoltaic energy systems. Springer
Lacasa D, Berenguel M, Yebra L, Martinez D (2006) Copper sintering in a solar furnace through fuzzy control. In: Proceedings of 2006 IEEE international conference on control applications, Germany
Lemos JM, Rato LM, Mosca E (2000) Integrating predictive and switching control: basic concepts and an experimental case study. In: Allgöwer F, Zheng A (eds) Nonlinear model predictive control. Birkhäuser Verlag, Basel, pp 181–190
Lemos JM (2006) Adaptive control of distributed collector solar fields. Int J Syst Sci 37(8):523–533
Meaburn A, Hughes FM (1993) Resonance characteristics of a distributed solar collector fields. Solar Energy 51(3):215–221
Meaburn A, Hughes FM (1994) Prescheduled adaptive control scheme for ressonance cancellation of a distributed solar collector field. Solar Energy 52(2):155–166
Miller FP, Vandome AF, McBrewster J (2010) Concentrating solar power. Alphascript Publishing
Murray-Smith R, Johansen TA (1997) Multiple model approaches to modelling and control. Taylor & Francis
Normey-Rico J, Bordons C, Berenguel M, Camacho EF (1998) A robust adaptive dead-time compensator with application to a solar collector field. In: Preprints 1st IFAC international workshop in linear time delay systems, New York, Pergamon, pp 105–110
Orbach A, Rorres C, Fischl R (1981) Optimal control of a solar collector loop using a distributed-lumped model. Automatica 27(3):535–539
Ordys AW, Pike MA, Johnson MA, Katebi, RM, Grimble, MJ (1994) Modelling and simulation of power generation plants. Springer, London
Paradkar, A, Davari A, Feliachi A (2002) Temperature control of a solarfurnace with disturbance accommodating controller. In: Proceedings of 34th southeastern symposium on system theory

Pasamontes M, Álvarez JD, Guzmán JL, Lemos JM, Berenguel M (2011) A switching control strategy applied to a solar collector field. Control Eng Pract 19:135145
Pereira C, Dourado A (2002) Application of neuro-fuzzy network with support vector learning to a solar power plant. In: Proceedings of 15th IFAC World Congress
Pickardt R (2000) Adaptive control of a solar power plant using a multi-model control. IEE Proc Theory Appl 147(5):493–500
Pickhardt R (2000) Nonlinear modelling and adaptive predictive control of a solar power plant. Control Eng Pract 8(8):937–947
Rato LM, Silva RN, Lemos JM, Coito F (1997) Multirate MUSMAR cascade control of a distributed colar field. In: Proceedings of the European control conference, ECC97, Brussells, Belgium
Rekioua D, Matagne E (2012) Optimization of photovoltaic power systems: modelisation, simulation and control. Springer
Roca L, Guzman J, Normey-Rico J, Berenguel M, Yebra L (2009) Robust constrained predictive feedback linearization controller in a solar desalination plant controller field. Control Eng Pract 17:1076–1088
Rorres C, Orbach A, Fischl R (1980) Optimal and suboptimal control policies for a solar collector system. IEEE Trans Autom Control AC-25 6:1085–1091
Rubio FR, Berenguel M, Camacho EF (1995) Fuzzy logic control of a solar power plant. IEEE Trans Fuzzy Syst 3(4):459–468
Silva RN, Filatov N, Hunbehauen H, Lemos JM (2005) A dual approach to start-up of an adaptive predictive controller. IEEE Trans Control Syst Technol 13(6):877–883
Silva RN, Rato LM, Lemos JM, Coito FV (1997) Cascade control of a distributed collector solar field. J Process Control 7(2):111–117
Silva RN, Lemos JM, Rato LM (2003a) Variable sampling adaptive control of a distributed collector solar field. IEEE Trans Control Syst Technol 11(5):765–772
Silva RN, Rato LM, Lemos JM (2003b) Time Scaling internal predictive control of a solar plant. Control Eng Pract 11(12):1459–1467
Soares AO, Gonçalves A, Silva, RN, Lemos JM (1997) A methodology for impact evaluation of alternative control strategies in a large scale power plant. Control Eng Pract 5(3):325–335
Stephanopoulos G (1984) Chemical process control: an introduction to theory and practice. Prentice-Hall (Int. Editions)
Tiwari GN (2002) Solar energy. Alpha Science International ltd, Pangborne
Valenzuela L, Zarza E, Berenguel M, Camacho EF (2004) Direct steam generation in solar boilers. IEEE Control Syst Mag 24(2):15–29
Valenzuela L, Zarza E, Berenguel M, Camacho EF (2005) Control concepts for direct steam generation in parabolic throughs. Solar Energy 78:301–311
Valenzuela L, Zarza E, Berenguel M, Camacho EF (2006) Control scheme for direct steam generation in parabolic throughs under recirculation operation mode. Solar Energy 80:1–17

Pasamontes M, Álvarez JD, Guzmán JL, Lemos JM, Berenguel M (2011) A switching control strategy applied to a solar collector field. Control Eng Pract 19:135–145

Pereira C, Dourado A (2002) Application of neuro-fuzzy network with support vector learning to a solar power plant. In: Proceedings of 15th IFAC World Congress, [illegible]

Pickardt R (2000) Adaptive control of a solar power plant using a multi-model control. IEE Proc Theory Appl [illegible]

[illegible] Nonlinear modelling and adaptive predictive control of a solar power plant [illegible]

Chapter 2
Models and Dynamics

This chapter describes models of distributed collector solar fields (DCSF) and uses them to discuss the typical dynamics found in these plants. The aim is not to describe a detailed model of a specific DCSF to be used for simulation purposes, as found in Camacho et al. (1988), since this would diverge from the objectives of the book. Instead, the attention is concentrated on the general dominant dynamics of DCSF plants in order to understand its structure and develop models that are adequate for adaptive control, both for algorithm design and to understand the difficulties that are to be faced by controller design.

Hence, we start by introducing a reduced complexity model that uses an energy balance to capture the spatial dependency of the system dynamics (with "space" meaning the position along the pipe) as well as on time. This model amounts to a hyperbolic partial differential equation whose solution is studied using mathematical methods and a heuristic discussion that relies on geometric arguments.

In order to prepare the ground for control algorithm development in subsequent chapters, the above infinite dimensional model is approximated to eliminate the dependency on the spatial dimension. These approximate models are an integral discrete time model obtained from the exact mathematical solution of the partial differential equation model, and continuous time state-space models obtained by projecting the spatial dependency on finite dimensional sets, either using finite differences or the orthogonal collocation method. Figure 2.1 summarizes these interdependencies between models.

The chapter concludes by making a parallel with various plants with similar dynamics, such as moisture control or glass tube manufacturing. The objective is to point out that the ideas and methods addressed throughout the book may also be applied in other types of plants.

J. M. Lemos et al., *Adaptive Control of Solar Energy Collector Systems*,
Advances in Industrial Control, DOI: 10.1007/978-3-319-06853-4_2,

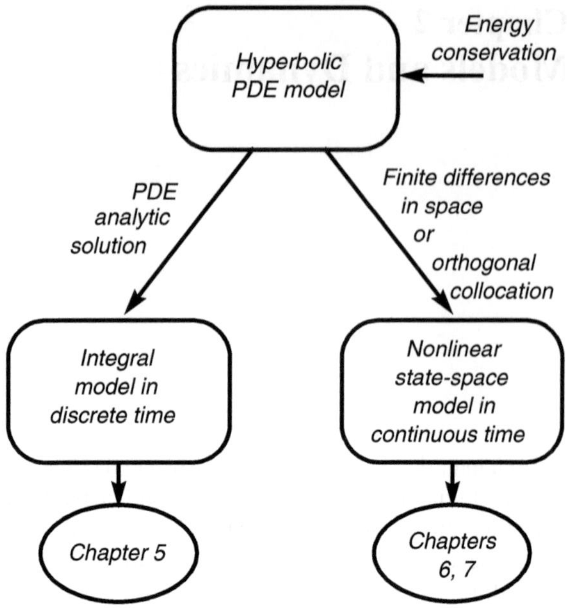

Fig. 2.1 DCSF model interdependency and the corresponding use for control design in subsequent chapters

2.1 Reduced Complexity Model

As explained in Chap. 1, the core of a DCSF is a metallic pipe inside which a working fluid, used to store energy, flows. Although the metallic pipe is inside a co-axial glass pipe used to enhance thermal efficiency by creating a greenhouse effect, this other element will not be considered in a first approach. In addition, it is assumed that there is no dependency of the fluid temperature on the radial dimension. Hence, the temperature of the fluid is described by a scalar function $T(z, t)$ where $z \in \mathbb{R}$ is a space dimension measured along the pipe, with $z = 0$ corresponding to the beginning of the pipe and $z = L$ to its maximum length, and $t \in \mathbb{R}$ is continuous time. Furthermore, it is assumed that T is the increment with respect to the environment temperature, so that $T = 0$ corresponds to the environment temperature.

In order to obtain an equation for the fluid temperature, consider the small section of the pipe between z and $z + \Delta z$, shown in Fig. 2.2. The net enthalpy accumulated in this pipe element between the time instants t and $t + \Delta t$ is given by

$$E_1 = \rho_f c_f A_f \Delta z \left[T(z, t + \Delta t) - T(z, t)\right],$$

where it is assumed that both Δt and Δz are small, ρ_f is the fluid density, c_f is the fluid specific heat, and A_f is the area of the pipe cross-section.

On the other hand, this net enthalpy is also given by the sum of two terms. One is the difference between the enthalpy entering and leaving the pipe element due to

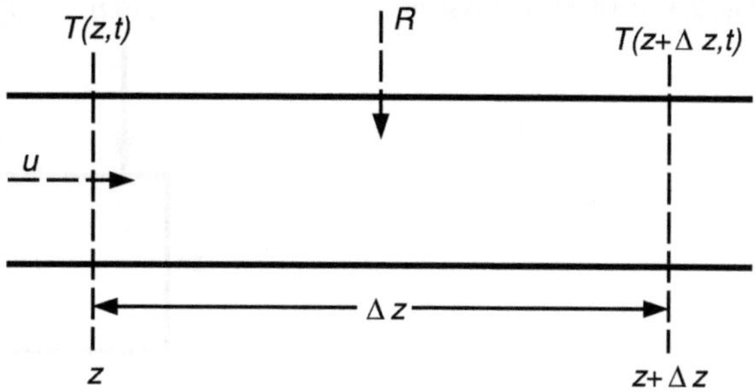

Fig. 2.2 An element of the pipe in a distributed collector solar field, and the variables used to deduce a PDE model

fluid flow, between t and $t + \Delta t$, which is given by

$$E_2 = \rho_f c_f \bar{u}(t)\Delta t \left[T(z,t) - T(z+\Delta z, t)\right],$$

where $\bar{u}$ is fluid flow. The other is the enthalpy increase due to solar energy, given by

$$E_3 = \bar{\alpha} R(t)\Delta z \Delta t,$$

where R is the solar radiation power and $\bar{\alpha}$ is a parameter related to the efficiency of energy absorption by the fluid, which depends on mirror optical efficiency and also on fluid thermal characteristics. Hence,

$$E_1 = E_2 + E_3$$

or

$$\rho_f c_f A_f \Delta z \left[T(z, t+\Delta t) - T(z,t)\right] = \rho_f c_f \bar{u}(t)\Delta t \left[T(z,t) - T(z+\Delta z,t)\right] + \bar{\alpha} R(t)\Delta z \Delta t. \tag{2.1}$$

Dividing now (2.1) by $\Delta z \Delta t$, observe that the fluid velocity u is given by

$$u = \frac{\bar{u}}{A_f}, \tag{2.2}$$

define

$$\alpha \triangleq \frac{\bar{\alpha}}{\rho_f c_f A_f}, \tag{2.3}$$

and let $\Delta t \to 0$ and $\Delta z \to 0$ to obtain the model expressed by the following partial differential equation

$$\frac{\partial}{\partial t} T(z,t) = -u \frac{\partial}{\partial z} T(z,t) + \alpha R(t). \tag{2.4}$$

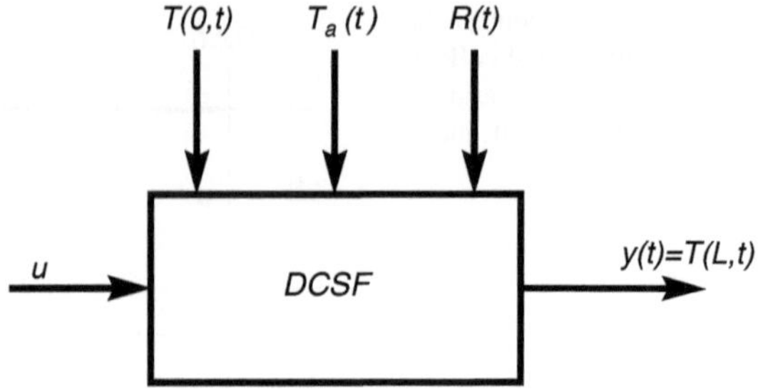

Fig. 2.3 A DCSF as a system

If heat losses to the environment are taken into account, an extra term appears and the model is given by the equation

$$\frac{\partial}{\partial t}T(z,t) = -u\frac{\partial}{\partial z}T(z,t) + \alpha R(t) - \gamma T(z,t), \tag{2.5}$$

γ is a loss coefficient. Due to the greenhouse effect created by the enveloping glass tube, the term corresponding to losses is in general less significant than the other terms. Equation (2.4) or, alternatively, (2.5), together with appropriate initial and boundary conditions, describes the dominant dynamics of a DCSF.

It is remarked that a detailed model of a DCSF build for simulation purposes must include other effects. These additions include the fact that the greenhouse effect causes an extra delay on the action of the solar radiation on the pipe. Furthermore, the specific heat c_f has a nonlinear dependence on the fluid temperature T that causes parameter α to depend on the operating point. Instead of modeling these effects, the control algorithms to follow use an approach based on parameter estimation. Furthermore, model (2.4) provides information on plant structure that is used in some of the algorithms.

2.2 Understanding the Dynamics

Figure 2.3 provides a view of a DCSF as a system. The manipulated variable input u is the fluid velocity (or, alternatively, the fuel flow), and the output $y = T(L,t)$ is the fluid temperature at the pipe outlet. There are three main disturbances, namely the fluid temperature at the pipe inlet, $T(0,t)$, the ambient temperature T_a, and the solar radiation $R(t)$. Of these disturbances, the solar radiation is the most significant due to the fast stresses it may cause, as well as for its strong and fast influence on fluid temperature. These variables are related by Eq. (2.5) or, with a further simplification, by (2.4).

The structure of the solutions of (2.4) and (2.5) is now examined. First, an exact solution is presented. Then, a qualitative discussion is made to gain insight into the type of response.

2.2.1 Exact Solution of the Distributed Model

The solution of Eq. (2.5) is defined in the plane $[z, t]$, in the set defined by $0 \leq z \leq L$ and $t \geq t_o$, where L is the length of the pipe and t_0 is the time instant in which plant operation starts. For solving (2.5), appropriate initial and boundary conditions must be defined. These conditions are the initial temperature distribution along the pipe, $T(z, t_0)$, obtained by fixing the time at the initial instant t_0, and the temperature of the fluid that enters the pipe as time passes, $T(0, t)$.

It is shown in Appendix B that the solution of Eq. (2.5) with the above initial and boundary conditions is given by

$$T(z, t) = T\left(z - \int_{t_0}^{t} u(\sigma)\,\mathrm{d}\sigma, t_0\right) e^{-\gamma(t-t_0)} + \alpha \int_{t_0}^{t} R(\sigma) e^{-\gamma(\sigma-t)}\,\mathrm{d}\sigma. \tag{2.6}$$

At this point, the reader may digress to Appendix A in order to deduce (2.6). Alternatively, the reader may proceed to follow a discussion about the properties of the solution (2.6) in Sects. 2.2.2 and 2.2.3, and then read Appendix A at a later time.

2.2.2 Qualitative Discussion

Figure 2.4 shows the plane $[z, t]$, the initial temperature distribution $T(z, t_0)$, the time evolution of the inlet fluid temperature $T(0, t)$ that is a boundary condition, and the temperature distribution along the pipe at an arbitrary time instant $t = t_1$, given by the function $T(z, t_1)$. By making the mute variable $t = t_1$, the solution formula (2.6) allows the computation of $T(z, t_1)$ as a function of $T(z, t_0)$ and $T(0, t)$.

Physical intuition tells that $T(z, t_1)$ is computed by making a shift to the right along the z-axis of the initial distribution $T(z, t_0)$ in a way that depends on the fluid flow. The left part of the pipe is filled with fluid entering through the inlet and this defines a temperature profile in this pipe section that depends on $T(0, t)$. In addition to this effect, there is a rise in temperature due to incident solar radiation, that acts in the same way in all the pipe, and that depends on the difference between t_1 and t_0. For t_0 constant, the larger the t_1 the more the radiation absorbed and the larger the impact on temperature increase.

Finally, there are losses to the environment. In the absence of solar radiation, $T(z, t_1)$ would be given not only by a shift of $T(z, t_0)$, but also by an attenuation due to

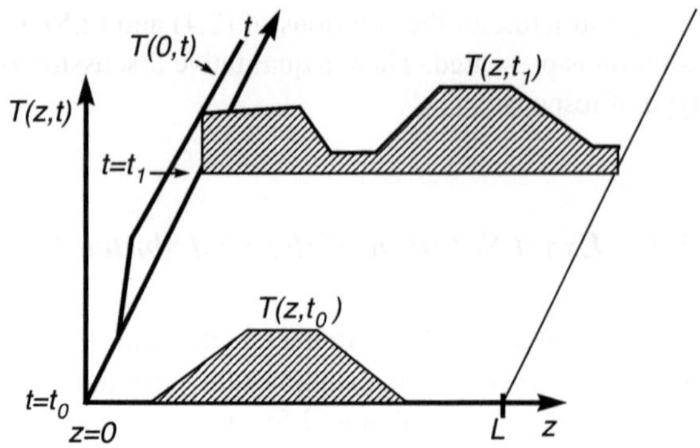

Fig. 2.4 Solution space of a DCSF with initial conditions

losses. Furthermore, when computing the cumulative effect of radiation there is also a discount factor due to losses that affects more the radiation received a longer time ago.

This physical intuition is actually confirmed by the solution (2.6). This equation computes the pipe distribution temperature $T(z, t)$ at an arbitrary instant t as the sum of two terms. The first term is

$$S_1 = T\left(z - \int_{t_0}^{t} u(\sigma)\, d\sigma, t_0\right) e^{-\gamma(t-t_0)}. \tag{2.7}$$

The meaning of S_1 is the following: To compute the temperature at position z and time t, pick-up the initial distribution $T(z, t_0)$ (that is assumed to be known), shift it to the right by $\int_{t_0}^{t} u(\sigma)\, d\sigma$ along the z-axis, and then attenuate it by $e^{-\gamma(t-t_0)}$, a factor that is less than 1 fot $t > t_0$.

The second term is

$$S_2 = \alpha \int_{t_0}^{t} R(\sigma) e^{-\gamma(\sigma-t)}\, d\sigma \tag{2.8}$$

and its effect is to raise the fluid temperature proportionally to the integral of the radiation discounted by a factor that depends on losses. The proportionality constant α depends, among other things, on mirror efficiency.

2.2.3 Characteristic Curves

Insight into DCSF dynamics may also be obtained by examining the special case when there are neither losses nor solar radiation, in which case (2.5) reduces to

$$\frac{\partial}{\partial t} T(z, t) = -u \frac{\partial}{\partial z} T(z, t). \tag{2.9}$$

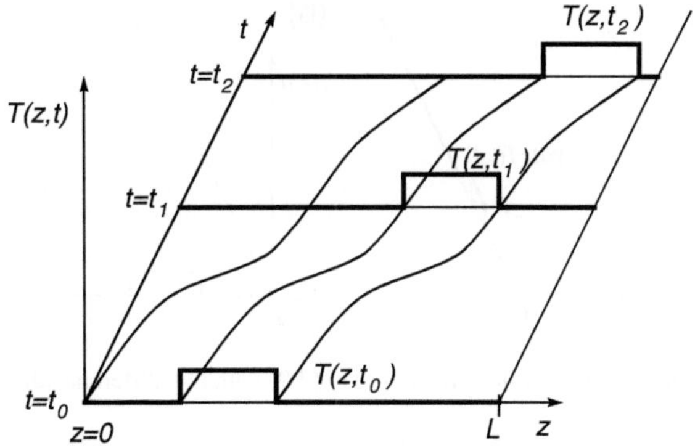

Fig. 2.5 Shift along characteristic curves of the solution of (2.9)

It is assumed in addition that the temperature of the fluid entering the pipe is $T(0, t) = 0$ at all times.

In the plane $[z, t]$ consider the curves whose coordinate z is given as a function of t by the solution of the differential equation

$$\frac{\mathrm{d}z}{\mathrm{d}t} = u \tag{2.10}$$

for various initial conditions on z. Along these curves, the solution $T(z, t)$ is constant. Indeed, along the solutions of (2.10), the total derivative of T with respect to time is, upon use of (2.10) and (2.9),

$$\frac{\mathrm{d}T}{\mathrm{d}t} = \frac{\partial T}{\partial z} \cdot \frac{\mathrm{d}z}{\mathrm{d}t} + \frac{\partial T}{\partial t} = u\frac{\partial T}{\partial z} + \frac{\partial T}{\partial t} = 0 \tag{2.11}$$

and hence it vanishes. For this reason, the solutions of (2.10) are called characteristic lines or curves and, as shown in Appendix B, they play a major role in the solution of (2.9) and also (2.5). Figure 2.5 shows an example of the solution of (2.9) obtained by propagating the initial temperature distribution along the characteristic lines.

If the flow u is constant, the solution of (2.10) is

$$z(t) = z(t_0) + u(t - t_0) \tag{2.12}$$

and the characteristic lines are straight lines. Figure 2.6 shows two situations in that the flow is constant but with different values in each case. The higher the flow the faster the temperature distribution along the pipe that affects the pipe outlet.

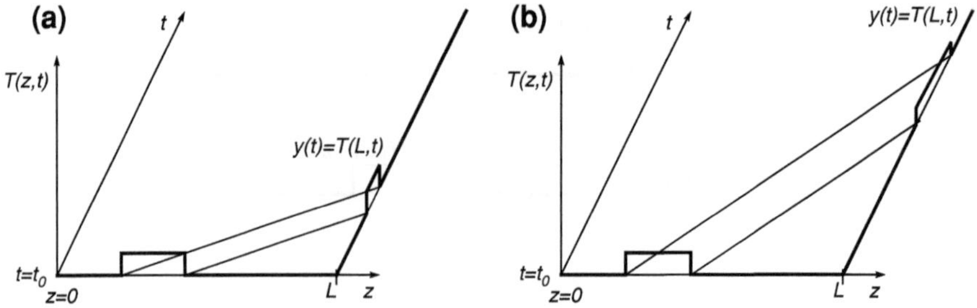

Fig. 2.6 **a** Bigger flow. **b** Smaller flow. Response of (2.9) with two different values of constant flow

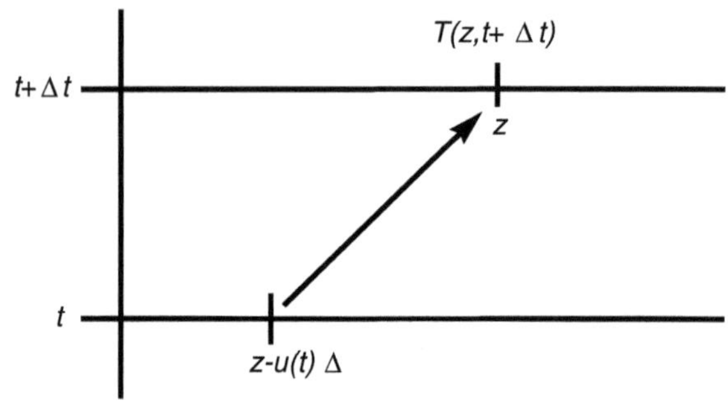

Fig. 2.7 Computing $T(z, t + \Delta t)$ from $T(z, t)$ using (2.13)

2.3 Discrete Time Integral Model

In order to obtain a discrete time integral[1] model, consider two time instants, t and $t + \Delta t$ and use the solution (2.6) to relate the respective temperature distributions. By approximating each integral in (2.6) by a rectangle, this gives

$$T(z, t + \Delta t) = T(z - u(t)\Delta t, t)e^{-\gamma \Delta t} + \alpha R(t)\Delta t \tag{2.13}$$

Equation (2.13) provides an approximate way to compute the temperature distribution at time $t + \Delta t$ from a previously known temperature distribution at time t. As shown in Fig. 2.7, to compute the temperature at position z, at time $t + \Delta t$, one should take the value of temperature at position $z - u(t)\Delta t$, at time t, multiply it by the attenuation due to losses, and then add the effect of radiation. This computation reflects the fact that the fluid is moving to the right with a velocity that, in the time interval $[t, t + \Delta t]$, is approximately $u(t)$ and, therefore, the fluid particle that was at position $z - u(t)\Delta t$ at time t moves forward to the right by a distance of $u(t)\Delta t$. Iterating (2.13) may be used to generate predictive models that are useful for control design, an approach that is explored in Chap. 5.

[1] The model is named "integral" in the sense that it is based in the integral of the PDE (2.10).

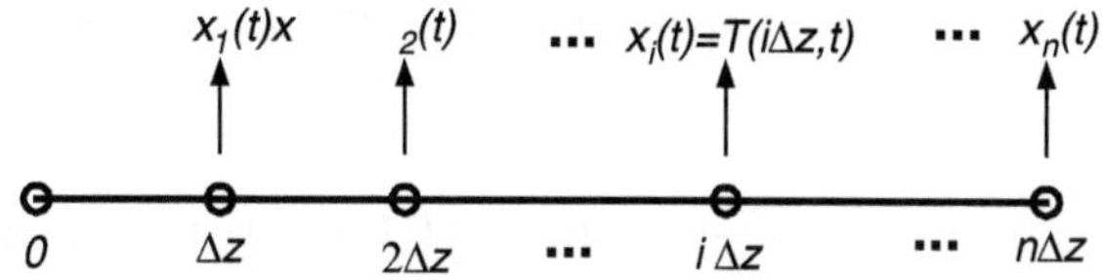

Fig. 2.8 Space grid to obtain a finite dimension state-space model of a DCSF

In a practical computation, the temperature is represented by its values in a space grid. In general, for computing $T(z, t + \Delta t)$, while z is in a grid point, $z - u(t)\Delta t$ is not, and hence $T(z - u(t)\Delta t)$ is not in the computer memory from the last iteration. One possibility to obtain $T(z - u(t)\Delta t)$ is then to approximate this quantity by interpolating the values of T at time t in neighboring points. Another possibility consists of using a varying time step Δt such that $u(t)\Delta t$ is constant and equal to the spatial grid increment Δz. As shown in Chap. 5, this option leads to predictive control algorithms with a variable time step that have the advantage of allowing fast setpoint changes without overshoot.

2.4 Finite Dimension State-Space Models

Instead of considering the fluid temperature in all the points of the pipe that lead to an infinite dimensional model, in this section we seek models in a finite dimensional state variable that is propagated in time, and from which an approximation of the temperature distribution can be recovered using algebraic operations.

2.4.1 Finite Difference Approach

One possibility is to take as state the temperature of equidistant points along the pipe. As shown in Fig. 2.8, let the points be located at $z = 0, z = \Delta z, \ldots, z = n\Delta z$, where n is the total number of points in the grid. For $i = 1, \ldots, n$, the state variable $x_i(t)$ is defined as the fluid temperature at position $i\Delta z$, according to

$$x_i(t) \triangleq T(i\Delta z, t). \tag{2.14}$$

Using this definition, and approximating the partial derivative with respect to z using backward finite differences, by

$$\frac{\partial}{\partial z}T(z, t) \approx \frac{T(z, t) - T(z - \Delta z, t)}{\Delta z}.$$

Equation (2.5) is approximated by the set of ordinary differential equations, for $i = 2, \ldots, n$

$$\frac{dx_i}{dt} = -u(t)\frac{1}{\Delta z}((1+\gamma)x_i(t) - x_{i-1}(t)) + \alpha R(t). \tag{2.15}$$

In the situations considered, the temperature measured for feedback purposes is the one of the pipe outlet. Hence, the output equation is

$$y(t) = x_n(t) \tag{2.16}$$

or

$$y(t) = Cx(t), \tag{2.17}$$

where

$$C = [1\,0 \ldots 0]$$

and y denotes the output variable in a system framework.

For $i = 1$, Eq. (2.15) reads as

$$\frac{dx_1}{dt} = -u(t)\frac{1}{\Delta z}((1+\gamma)x_1(t) - T(0,t)) + \alpha R(t). \tag{2.18}$$

The inlet fluid temperature $T(0, t)$ enters Eq. (2.18) as an external signal that corresponds to a disturbance.

Defining the state vector x as the temperature along the grid points,

$$x \triangleq [x_1 \ldots x_n],$$

and the parameter matrices

$$\Xi \triangleq [1 \ldots 1]^T,$$

$$\vec{e}_1 = [1\,0 \ldots 0]^T,$$

and

$$B = -\frac{1}{\Delta z}\begin{bmatrix} 1+\gamma & 0 & \ldots & 0 \\ -1 & 1+\gamma & \ddots & \vdots \\ \vdots & \ddots & \ddots & 0 \\ 0 & \ldots & -1 & 1+\gamma \end{bmatrix}$$

the set of Eqs. (2.15), (2.18) may be written in compact form as

$$\dot{x} = Bxu + \alpha \Xi R + \overrightarrow{e}_1 \frac{u}{\Delta z} T(0, t). \tag{2.19}$$

It is remarked that matrix B is invertible and has n eigenvalues equal to $-(1+\gamma)$.

2.4.2 Reachable States

In a dynamical system such as a DCSF, the reachable states are the plant model states that can be attained, starting from the origin by manipulating the plant input within its admissible values. Characterizing the set of reachable states is of interest because it provides the ground for the control design methods based on feedback linearization that are considered in Chap. 6. In addition, this characterization shows the states that can be attained, which is important information. Indeed, the discussion made hereafter shows that, in a DCSF, when starting from a state corresponding to all fluid elements and the inlet fluid at the ambient temperature, the only possible temperature distributions along the pipe are monotonically increasing functions of space. Furthermore, for bounded fluid flow cases (as is always the case in a DCSF), the fluid temperature along the pype is also bounded.

Using methods from nonlinear dynamical systems (Nijmeijer and van der Schaft 1990; Barão et al. 2002), it is possible to show that the only directions along which the state x of the nonlinear lumped parameter model (2.19) may change are linear combinations of the three following vectors:

$$\begin{bmatrix} 1 \\ 1 \\ \vdots \\ 1 \end{bmatrix}, \begin{bmatrix} 1 \\ 0 \\ \vdots \\ 0 \end{bmatrix}, Bx.$$

The first vector concerns heating by solar radiation. Under constant radiation the fluid is heated uniformly along the pipe, an event that corresponds to a state trajectory aligned with the direction $\begin{bmatrix} 1 & 1 & \dots & 1 \end{bmatrix}^T$.

The second vector corresponds to the possibility of manipulating the temperature derivative with respect to space at the pipe inlet.

Finally, the third vector represents fluid transport along the pipe.

2.4.3 Jacobian Linearization

To understand the dynamics of the linearized system around an equilibrium point, assume that the radiation R is constant and that the temperature of the inlet fluid is zero, $T(0, t) = 0$. Let u_{eq} be a constant velocity of the fluid and let x_{eq} denote the corresponding value of the state at equilibrium. Both these variables satisfy

$$Bx_{eq}u_{eq} + \alpha \Xi R = 0. \tag{2.20}$$

Since B is invertible and u_{eq} is a scalar, the distribution of temperature at equilibrium can be computed explicitly by

$$x_{eq} = \frac{\alpha R}{u_{eq}} B^{-1} \Xi. \tag{2.21}$$

Let a perturbation $\Delta u(t)$ be applied to the equilibrium value. The state will respond with a variation $\Delta x(t)$ around its equilibrium x_{eq}, as well as the output y. The input and state perturbations are related by the linear matrix ODE

$$\frac{d}{dt}\Delta x = \left.\frac{\partial \phi}{\partial x}\right|_{\substack{x = x_{eq} \\ u = u_{eq}}} \cdot \Delta x + \left.\frac{\partial \phi}{\partial u}\right|_{\substack{x = x_{eq} \\ u = u_{eq}}} \cdot \Delta u, \tag{2.22}$$

where

$$\phi(x, u) = BXu + \alpha \Xi R. \tag{2.23}$$

Thus, the linearized model is

$$\frac{d}{dt}\Delta x = Bu_{eq}\Delta x + Bx_{eq}\Delta u \tag{2.24}$$

Since the output variable y is related to the state by (2.17), the corresponding increments are also related by

$$\Delta y = C\Delta x. \tag{2.25}$$

The static gain of the incremental model is obtained from (2.24) by equating the derivative to zero, yielding $-Cx_{eq}/u_{eq}$. The fact that the static gain is negative is readily interpreted in physical terms. When the flow increases, the residence time of the fluid particles inside the pipe decreases, as well as their temperature when they reach the outlet, because the particles receive less radiation.

The transfer function relating Δu with Δy is obtained by taking the Laplace transform with zero initial conditions of both sides of (2.24) and using (2.25). Observing that

$$\det(sI - Bu_{eq}) = (s + (1 + \gamma)u_{eq})^n, \tag{2.26}$$

the transfer function of the linearized system is

$$H(s) = \frac{C\,\mathrm{adj}(sI - Bu_{eq})B}{(s + (1 + \gamma)u_{eq})^n}, \tag{2.27}$$

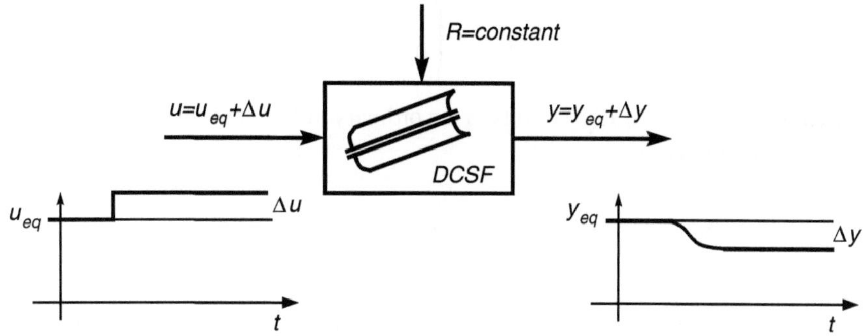

Fig. 2.9 Disturbance of an equilibrium in a DCSF

where adj denotes the adjoint of a matrix. Thus, $H(s)$ has n real and negative poles that are located at $-(1+\gamma)u_{\text{eq}}$. Since the poles are proportional to the equilibrium flow, it is concluded that the incremental response becomes faster when the flow increases. Figure 2.9 summarizes the discussion about the linearized system.

2.4.4 Orthogonal Collocation

Instead of relying on the approximation of derivatives by finite differences, the orthogonal collocation method (OCM) yields a finite dimensional approximation to the solar collector field model (2.5) by projecting the function corresponding to the exact solution $T(z,t)$ in a finite set of functions and obtaining a set of ordinary differential equations for the coefficients that express $T(z,t)$ in the basis functions (Villadsen and Michelsen 1978). This is a method commonly used in tubular and other type of chemical reactors (Dochain et al. 1992; Rice and Do 1995) that may also be used for DCSFs, with the advantage of yielding lower dimensional models than finite differences.

In order to apply the OCM to approximate (2.5) (for simplicity losses are assumed to be neglected) by a set of ordinary differential equations, it is assumed that the temperature along the pipe $T(z,t)$ is represented by the weighted sum

$$T(z,t) = \sum_{i=0}^{N+1} \varphi_i(z)\, T_i(t), \tag{2.28}$$

where $T_i(t)$ are time weights that define the state to be propagated and the functions $\varphi_i(z)$ are Lagrange interpolation polynomials, orthogonal at the so-called interior collocation points z_i for $i = 1, \ldots, N$ and at the boundary collocation points z_0 and z_{N+1}. The polynomials $\varphi_i(z)$ verify thus at the collocation points

$$\varphi_i(z_j) = \begin{cases} 1 & i = j \\ 0 & i \neq j \end{cases}. \tag{2.29}$$

Inserting (2.28) into Eq. (2.5) results in the ordinary differential equation verified by the time weights $T_i(t)$

$$\sum_{i=0}^{N+1} \varphi_i(z) \frac{\mathrm{d}T_i(t)}{\mathrm{d}t} = -\frac{u}{L} \sum_{i=0}^{N+1} \frac{\mathrm{d}\varphi_i(z)}{\mathrm{d}z} T_i(t) + \alpha\, R(t). \tag{2.30}$$

Compute now (2.30) at each of the collocation points $z = z_j$. Since (2.29) holds, and individuating the term $i = 0$ corresponding to the boundary conditions, it follows that

$$\frac{\mathrm{d}T_j(t)}{\mathrm{d}t} = -\frac{u}{L} \left[\sum_{i=1}^{N+1} \frac{\mathrm{d}\varphi_i(z_j)}{\mathrm{d}t} T_i(t) + \frac{\mathrm{d}\varphi_0(z_j)}{\mathrm{d}t} T_0(t) \right] + \alpha R(t). \tag{2.31}$$

By making $j = 1, \ldots, N+1$, that is to say, by considering all the collocation points apart from the first, the PDE (2.5) is therefore approximated by $n = N+1$ ordinary differential equations (ODE). The state of this nonlinear ODE system is formed of $T_i(t)$, which are the temperatures at the collocation points.

In matrix form, this lumped parameter model is written

$$\dot{x} = Bxu + \Upsilon x_0 u + \Xi \alpha R(t), \tag{2.32}$$

where

$$x = \begin{bmatrix} T_1 & T_2 & \cdots & T_{N+1} \end{bmatrix}^T, \tag{2.33}$$

is the state, with $T_i(t) = T(z_i, t)$, and z_i are the collocation points, the matrices B, Υ and Ξ are given by:

$$B = -\frac{1}{L} \begin{bmatrix} \varphi_1'(z_1) & \varphi_2'(z_1) & \cdots & \varphi_{N+1}'(z_1) \\ \varphi_1'(z_2) & \varphi_2'(z_2) & \cdots & \varphi_{N+1}'(z_2) \\ \vdots & \vdots & \ddots & \vdots \\ \varphi_1'(z_{N+1}) & \varphi_2'(z_{N+1}) & \cdots & \varphi_{N+1}'(z_{N+1}) \end{bmatrix}, \tag{2.34}$$

$$\Upsilon = -\frac{1}{L} \begin{bmatrix} \varphi_0'(z_1) \\ \varphi_0'(z_2) \\ \vdots \\ \varphi_0'(z_{N+1}) \end{bmatrix}, \qquad \Xi = \begin{bmatrix} 1 \\ 1 \\ \vdots \\ 1 \end{bmatrix}, \tag{2.35}$$

where

$$\varphi_j'(z_i) \triangleq \frac{d\varphi_j(z)}{dz}\,|_{z=z_i} \tag{2.36}$$

and $T_0(t)$ is the boundary condition.

Equation (2.32) has the same structure as the equation that describes the finite dimensional model obtained by finite differences and given by Eq. (2.19). The main difference consists in the fact that, for the same level of the approximation error, orthogonal collocation requires a state of much smaller order than the finite difference method.

2.5 Plants with Similar Dynamics

Most of the ideas on control design discussed in this book can be applied to plants with a dynamics that is similar to that of DCSFs. In broad terms, these plants are characterized by processes with both a temporal and spatial dependence in which a fluid flows along a scalar dimension (for instance inside a pipe) and with energy exchanges along the spatial dimension. The manipulated variable is either the fluid speed or flow, or the power received along the space dimension. Furthermore, it is assumed that diffusion is negligible.

It is remarked that, although in DCSFs the manipulated variable is usually taken to be the fluid flow, it is possible to make a change in variables such as to define a virtual manipulated variable that amounts to be a correction of incident solar radiation, as shown in Chap. 5.

In mathematical problems, the plants considered are described by a Cauchy problem (Pazy 1983) associated to a hyperbolic PDE like (2.41), where a variable $T(z, t)$ that is not necessarily the temperature, evolves in time from an initial condition $Tz, 0)$, with t denoting time and $0 \leq z \leq L$ denoting the scalar coordinate of the space dimension.

The examples described hereafter that illustrate this class of plants are:

- Moisture drying
- Traffic in highways
- Glass tube manufacturing
- Air heating fan
- Steam superheaters

These examples illustrate (but by no means exhaust) the variety of plants that can be controlled with the methods described in Chaps. 2–8. Furthermore, some of them will be used to illustrate briefly the use of these methods in a context outside the DCSF area.

In addition to the above examples, control of trailing centerline in arc welding is also described. Although this process is no longer described by an hyperbolic model because, in addition to movement, there are significant diffusion effects, controller design for it can be done using the algorithms of Chaps. 3 and 4. Furthermore, the

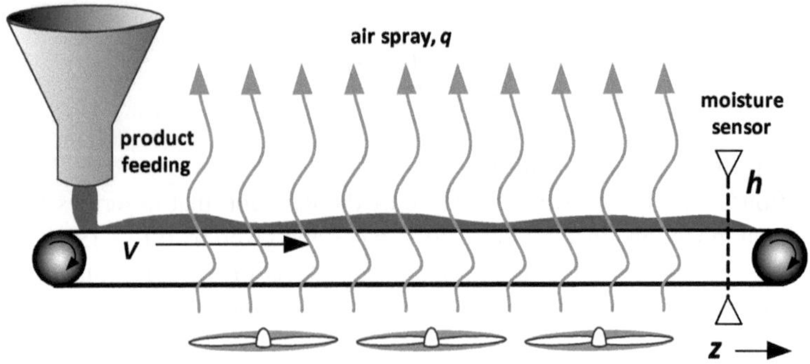

Fig. 2.10 Schematic view of a drying plant

key idea of Chap. 5 of using a sampling interval that depends on movement velocity may also be used with advantage.

2.5.1 Moisture Control

Drying is an operation frequently performed in several types of industrial processes such as those found in the pharmaceutical industry, the food industry (such as "Italian pasta manufacturing"), cereal processing, or coal preparation for use in thermoelectric power plant units. The objective is to reduce the moisture content of a given product, usually provided in the form of powder, small grains, or porous paste (Nybrant 1988; Moreira and Bakker-Arkema 1990).

Figure 2.10 shows a schematic view of a moisture drying line that consists of a conveyor belt that moves with a velocity v, and over which the material to dry is transported. At the beginning of the conveyor belt a feeder drops the product to dry in order to form a layer over the moving belt with an approximately constant thickness. In order to remove the moisture, a number of fans impel hot air through the belt with a flow $q(t)$.

Let $w(z, t)$ denote the moisture quantity per unit volume of the material spread along the belt, where z is the abscissa of a coordinate system aligned with the belt that has the origin at the feeding point and t is time. It is assumed that the rate of moisture removal is proportional to $q(t)w(z, t)$.

In order to obtain a mathematical model of this process consider a small section of the belt between the points with abscissa z and $z + \Delta z$, with Δz small. The difference between the moisture quantity inside the volume element at times t and $t + \Delta t$, where Δ is a small increment of time, is given by

$$M_1 = A_b \Delta z \left[w(z, t + \Delta t) - w(z, t) \right],$$

where A_b denotes the area of the cross section of the material over the belt.

On the other hand, the accumulation of moisture in the volume element due to the belt movement is given by

$$M_2 = A_b v \Delta t \left[w(z,t) - w(z+\Delta z, t) \right],$$

while there is a quantity of moisture leaving the element of volume that is given by

$$M_3 = \beta A_b w(z,t) \Delta z \Delta t,$$

where β is a parameter. A mass balance applied to the moisture results in

$$M_1 = M_2 + M3.$$

Dividing by $\Delta z \Delta t$ and making the increments $\Delta z \to 0$, $\Delta t \to 0$ results in the model

$$\frac{\partial}{\partial t} w(z,t) = -v \frac{\partial}{\partial z} w(z,t) - \beta q(t) w(z,t), \tag{2.37}$$

an equation similar to (2.6).

2.5.2 *Traffic in Highways*

Traffic-flow modeling and control is an area that attracts increasing attention and provides an example in which the process variable of interest is not temperature. A simple model of a unidirectional highway section with neither sinks nor sources is obtained as follows: Let $\rho(z,t)$ denote the concentration (number of vehicles per unit length) at position z of the highway and time t, and $q(z,t)$ denote the flow (number of vehicles that pass at point z per unit time. Consider an element of the highway between positions z and $z + \Delta z$ and during a time interval Δt, with both Δz and Δt small.

By the definition of ρ, the accumulation of vehicles inside the element during the time interval between t and $t + \Delta t$ is given by

$$N_1 = \Delta z(\rho(z, t + \Delta t) - \rho(z,t)).$$

On the other hand, this same quantity is also given by the number of vehicles that enter the highway element at the point with coordinate z, minus the vehicles that leave the highway element at $z + \Delta z$, that is to say

$$N_2 = \Delta t(q(z,t) - q(z + \Delta z, t)).$$

Equating N_1 and N_2, dividing by $\Delta z \Delta t$ and making the increments $\Delta z \to 0, \Delta t \to 0$ yields the conservation equation

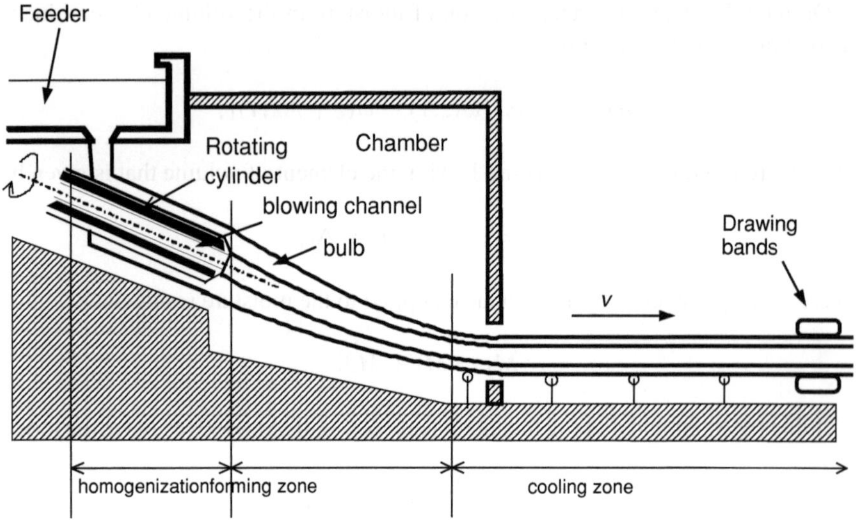

Fig. 2.11 Schematic view of the glass tube drawing bench. According to Wertz et al. (1987)

$$\frac{\partial}{\partial t}\rho(z,t) = -\frac{\partial}{\partial z}\rho(z,t). \tag{2.38}$$

Equation (2.38) can be expressed in terms of vehicle density ρ and vehicle speed u by using the relationship (Lieu 2011)

$$q(z,t) = \rho(z,t)u(z,t) \tag{2.39}$$

Using (2.39), and assuming that there are sinks/sources of vehicles, the traffic flow model reads as

$$\frac{\partial}{\partial t}\rho(z,t) = -\frac{\partial}{\partial z}\left(\rho(z,t)u(z,t)\right) + g(z,t), \tag{2.40}$$

where g is the generation dissipation of vehicles per unit length, per unit time.

2.5.3 Glass Tube Manufacturing

The glass tube drawing bench depicted in Fig. 2.11 works as follows: Molten glass drops from a feeder bowl over a tilted rotating cylinder. The air blown through an axial hole located inside the cylinder forms the tube from the glass bulb at the cylinder bottom.

Both the diameter of the tube and the thickness of its glass walls are associated to a dynamics that depends both on time and space, with the space dimension defined by

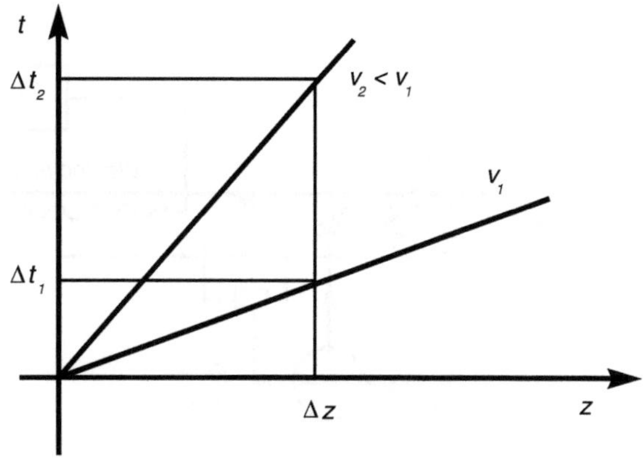

Fig. 2.12 Characteristic lines associated with glass tube movement in two different situations, corresponding to two tube constant velocities, v_1 and $v_2 < v_1$

the length along the tube. In the identification study performed in Wertz et al. (1987) the movement of the glass tube is taken into consideration by using a sampling interval that is inversely proportional to the tube velocity. This procedure can be understood by considering the simplified model of the glass movement given by

$$\frac{\partial}{\partial t} y(z, t) = -v \frac{\partial}{\partial z} y(z, t), \tag{2.41}$$

where $y(z, t)$ is the tube diameter at position z (measured along the tube) and at time t, and v is the tube velocity. Equation (2.41) is similar to Eq. (2.41) and does nothing more than to represent a shift to the right of the function $y(z, 0)$ that describes the initial tube diameter along the bench. As explained in Sect. 2.2.3, this shift is such that the function $y(z, t)$ is constant along the characteristic lines in the $[z, t]$ plane, given by the solutions of the differential equation

$$\frac{\mathrm{d}z}{\mathrm{d}t} = v. \tag{2.42}$$

For constant v the general solution of (2.42) is

$$z(t) = z(0) + vt, \tag{2.43}$$

meaning that the characteristic lines are straight lines of slope v. Thus, as shown in Fig. 2.12, a point at the origin of the glass tube, moving with velocity v_1, needs a time Δt_1 to reach a point of abcissa Δz, but needs a greater time Δt_2 to reach the same point if the velocity is smaller. This means that, if the sampling interval is smaller, and if it is chosen so as to be adequate for a given velocity, it might be too big if the velocity is bigger (since the details of the dynamic response are lost), or oversampling might occur if the velocity is smaller.

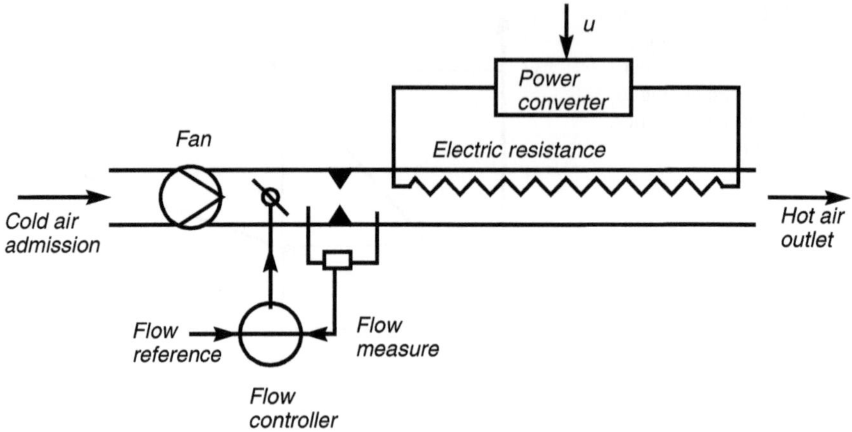

Fig. 2.13 Schematic view of the air heating fan system

Furthermore, although the transport delay of the continuous time model varies with the velocity, it becomes constant in a discrete model if the sampling time is inversely proportional to the velocity.

These observations suggest to consider the transformed timescale τ defined by

$$\tau = vt, \tag{2.44}$$

in which (2.41) reads

$$\frac{\partial}{\partial \tau} y(z, \tau) = -\frac{\partial}{\partial z} y(z, \tau), \tag{2.45}$$

The characteristic lines associated to the normalized Eq. (2.45) have a slope of 45°. This normalization is explored in Wertz et al. (1987) to obtain the same model for different tube sizes. The idea of making a normalization with respect to velocity is common to other areas such as control applications to aerospace engineering (Serra et al. 2010). This idea is also useful to design controllers for DCSFs (Silva et al. 2003a, b), as explained in Chap. 5.

2.5.4 Air Heating Fan

The air heating fan system shown in Fig. 2.13 has also a dynamical behavior that is similar to those of DCSF. This system consists of a tube inside which cold air is blown by a fan. The air flow can be adjusted with a register that is manipulated by a flow controller. Along the tube, an electric resistor liberates heat with a power that depends on the command signal (u in Fig. 2.13) of an electronic power converter. The temperature of the heated air is measured at the tube outlet.

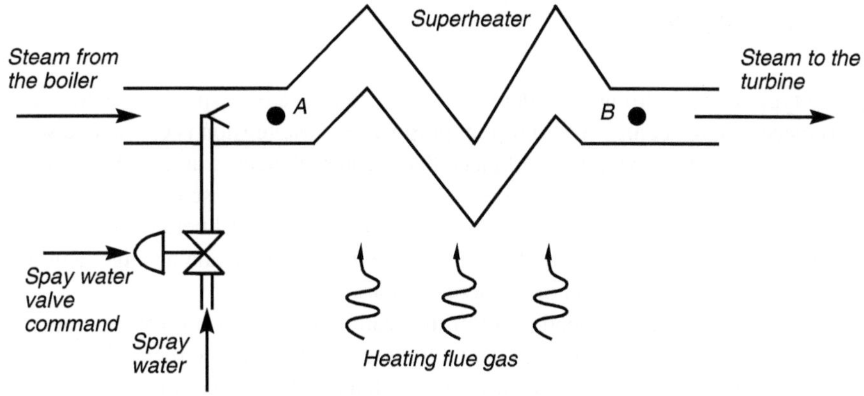

Fig. 2.14 Schematic view of the superheated steam temperature manipulation system

Neglecting diffusion, this pilot plant can be modeled by the PDE

$$\frac{\partial}{\partial t}T(z,t) = -f(t)\frac{\partial}{\partial z}T(z,t) + \alpha R(t), \tag{2.46}$$

where $T(z,t)$ is the air temperature at position z measured along the tube and at time t, $R(t)$ is the heating power and α is a parameter.

This plant is used in Chap. 4 to illustrate a control design procedure based on multiple models that is subsequently applied to a DCSF.

2.5.5 Steam Superheater

Boiler units for thermoelectric power production, either solar or with fossil fuel, include steam superheaters that have a twofold function: Eliminate the water droplets in the steam that could damage the turbine blades and increase energy use efficiency.

Associated to some superheaters there are spray water injection systems that provide a way to regulate the steam temperature at a required value (the higher as possible to maximize efficiency, but low enough to avoid damaging equipment).

Figure 2.14 shows a simplified schematic view of the so-called attemperator system, where the spray water valve is operated so as to keep the temperature of the steam leaving the superheater, at point B, close to the desired setpoint. The superheater consists of a long pipe, or parallel of pipes, inside which a fluid (steam) circulates to be heated by an external source. Up to a first approximation, the dynamics of the superheated is therefore similar to 2.46, with the velocity of response of the temperature at point B in response to a change in temperature at point A depending on steam flow.

Like the other examples in this chapter, superheated steam temperature can be controlled with the techniques described in subsequent chapters.

2.5.6 Arc Welding

In the type of arc welding considered here, the objective is to spread a seam over a workpiece (Santos et al. 2000). For that purpose, electric tension is created between a consumable electric and the workpiece that creates an electric arc. In this process, particles of the consumable electrode are liberated and form the seam over the workpiece. The consumable electrode passes through a torch that liberates a protective gas. Figure 2.15 shows a schematic view of the welding machine arrangement and Fig. 2.16 shows a photograph of the actual apparatus.

The temperature at the head point of the seam, where the arc forms, is not controllable because it corresponds to the melting temperature of the metal. Hence, the approach followed consists in measuring the temperature to be controlled 2.5 cm behind the point where the arc is formed. This means that what is controlled is the rate of cooling of the seam temperature, in what is referred in technical terms to be the trailing centerline temperature (Santos et al. 2000). This temperature is measured with a pyrometer. A screen protects the pyrometer from receiving direct radiation from the melting point. The whole set (torch, screen and pyrometer) is connected to a car that moves with a velocity that is optimized, depending on the type of materials involved and the electric tension, in order to obtain a stable arc. In the experiments reported in this book, the relation tension/car speed is optimized for the protection gas Ar–CO_2 20–80 % and the electrode made of carbon-steel ϕ 1.2 mm.

The process involves complex thermal-electro-chemical phenomena. The temperature in the workpiece can be modeled by the Fokker–Planck equation. A major difference with respect to DCSFs consists in the fact that, in arc welding, heat diffusion is much significant, resulting in a model of parabolic type. A snapshot of the space dependency of temperature at one instant of time is shown in Fig. 2.17. The shape of the isothermal curves is affected by several factors, including the car velocity and, most important, the geometric shape of the workpiece.

Hence, due to the complexity of the process it is no surprise that a high level of uncertainty affects modeling. When approximating the relationship between electric tension and temperature by a lumped parameter linear model, such as an ARX model, identified from process data, there is high variability in the model parameters when the data set changes. This is illustrated in Fig. 2.18 which shows frequency response curves corresponding to linear models identified for rectangular workpieces of the same steel, and all with the same rectangular shape, with widths of 6 and 12 mm. As it is apparent, even at low frequency there is strong variability. When considering different workpieces of different shapes and of different steels, variability is even greater and prevents the possibility of designing a single controller of constant gains to stabilize all possible plant outcomes with an acceptable performance. Thus, this example provides a good example where adaptive control may be applied with advantage.

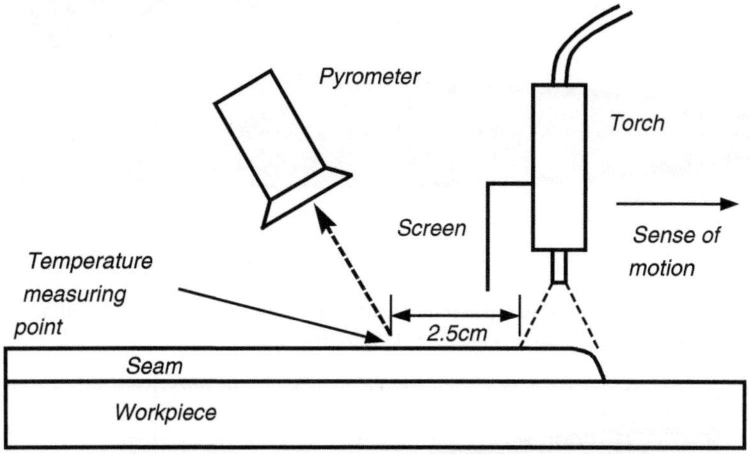

Fig. 2.15 A schematic view of the welding machine apparatus

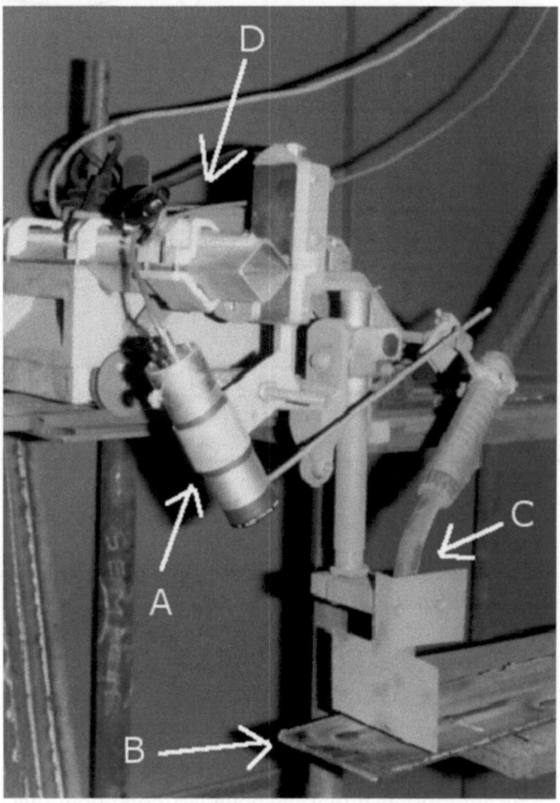

Fig. 2.16 A photograph of the welding machine setup. **a** Optical pyrometer, protected from direct radiation by a screen; **b** Plate to weld; **c** The torch, inside which passes an electrode; **d** The cart that makes the torch move

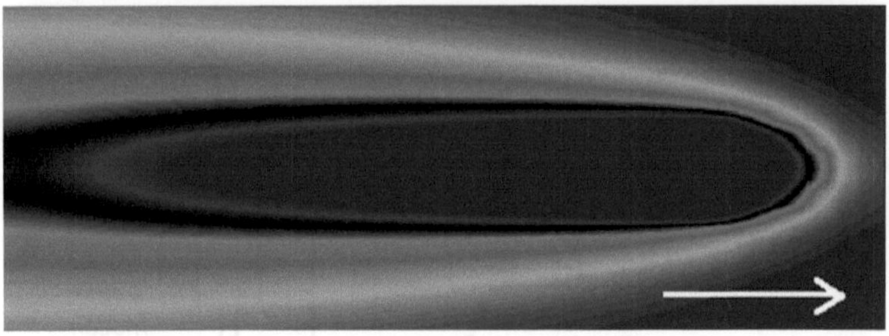

Fig. 2.17 A time snapshot of the spatial distribution of temperature in the arc welding example. The *arrow* indicates the direction of movement of the torch

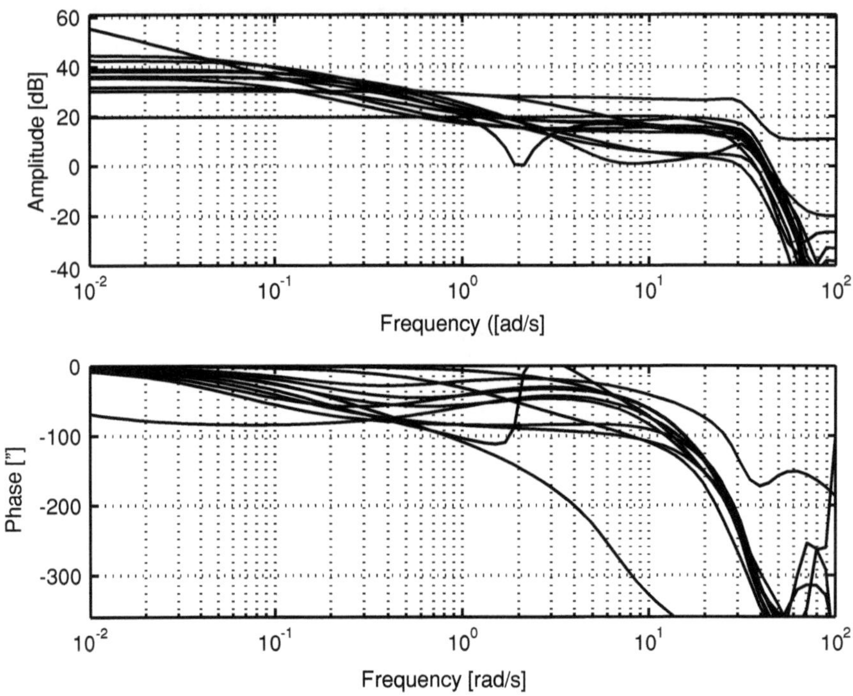

Fig. 2.18 Frequency response of transfer functions identified using different data sets between the electric tension and the temperature

2.6 Main Points of the Chapter

The dynamics of DCSF depends on space as well as on time and, as such, is represented by a PDE with appropriate initial and boundary conditions that reflect physical conditions, namely the initial distribution of fluid temperature along the pipe and fluid temperature at the inlet. The solution of this model reflects the fact that the fluid is moving to the right, that it receives energy from solar radiation which increases its temperature, and that it is subject to thermal losses to the environment, which tends to decrease its temperature. It is possible to obtain a closed-form solution of the model and to make a numerical approximation of it resulting in a recursive model that can be used to make predictions of temperature values. In addition to this model, nonlinear state-space models with a bilinear structure are obtained in this chapter. All the models are explored in subsequent chapters in order to develop different controllers. Attention is called to the fact that the dynamics found in DCSF are common to other, seemingly different, plants that include transport phenomena and that, therefore, the ideas explored here are also useful in other system control problems, with some examples being provided.

2.7 Bibliographic Notes

A detailed model of the PSA DCSF used, which is based on physical principles and parameter estimation from plant data, is described in Camacho et al. (1988, 1997). In addition to the energy balance inside the fluid that accumulates solar radiation, the model includes an equation obtained by making an energy balance in the metal of the pipe and a nonlinear function that represents the dependence of some coefficients on temperature. A detailed review of different models is presented in Camacho et al. (2007). Bilinear state-space models are presented in Barão et al. (2002) and input/output predictive models with variable sampling in Silva et al. (2003a, b).

The solution of PDE that model processes involving mass and/or energy transport phenomena are studied in many books, some examples of which are Jeffrey (2003), Parker (2003), Grindrod (1996). The Fokker–Plank equation, which is an important model for processes that involve diffusion as well as transport and which appears in wide different areas ranging from thermal processes to telecommunications, is studied in Grindrod (1996) by using an approach that combines intuition and sound mathematical basis.

The highway traffic-flow model described is an aggregated model in the spirit of Chapter 2 of Lieu (2011). An agent-based alternative is described in Kerner and Klenov (2003), Marques and Neves-Silva (2011).

References

Barão M, Lemos JM, Silva RN (2002) Reduced complexity adaptive nonlinear control of a distributed collector solar field. J Process Control 12:131–141

Camacho EF, Rubio FR, Gutierrez JA (1988) Modelling and simulation of a solar power plant with a distributed collectors system. In: Proceedings of IFAC symposium on power systems, modelling and control applications, Brussels, pp 11.3.1–11.3.5

Camacho EF, Berenguel M, Rubio F (1997) Advanced control of solar plants. Springer, New York

Camacho EF, Rubio FR, Berenguel M, Valenzuela L (2007) A survey on control schemes for distributed solar collector fields. Part I. Modeling and basic control approaches. Sol Energy 81:1240–1251

Dochain D, Babary JP, Tali-Maamar N (1992) Modelling and adaptive control of a nonlinear distributed parameter bioreactors via orthogonal collocation. Automatica 28(5):873–883

Grindrod P (1996) The theory and applications of reaction-diffusion equations, 2nd edn. Clarendon Press, Oxford

Jeffrey A (2003) Applied partial differential equations. Academic Press, New York

Kerner BS, Klenov SL (2003) A microscopic theory of spatial-temporal congested traffic patterns at highway bottlenecks. Phys Rev E 68(3):036130

Lieu H (2011) Traffic-flow theory. Public Roads 62(4) (US Department of Transportation)

Marques M, Neves-Silva R (2011) Development of a microscopic driver-vehicle model using a control theory approach. Int J Model Simul 31(3):210–217

Moreira RG, Bakker-Arkema FW (1990) A feedforward/feedback adaptive controller for commercial cross-flow grain driers. J Agric Eng Res 45:107–116

Nijmeijer H, van der Schaft AJ (1990) Nonlinear dynamical control systems. Springer, New York

Nybrant TG (1988) Modelling and adaptive control of continuous grain driers. J Agric Eng Res 40(3):165–173

Parker DF (2003) Fields, flows and waves. Springer, New York

Pazy A (1983) Semigroups of linear operators and applications in partial differential equations. Springer, New York

Rice GR, Do DD (1995) Applied mathematics and modeling for chemical engineers. Willey, Chichester

Santos TO, Caetano RB, Lemos JM, Coito FJ (2000) Multipredictive adaptive control of arc welding trailing centerline temperature. IEEE Trans Control Syst Technol 8(1):159–169

Serra P, le Bras F, Hamel T, Silvestre C, Cunha R (2010) Nonlinear IBVS controller for the flare maneuver of fixed-wing aircraft using optical flow. In: Proceedings of 49th IEEE conference on decision and control (CDC 2010), Atlanta

Silva RN, Lemos JM, Rato LM (2003a) Variable sampling adaptive control of a distributed collector solar field. IEEE Trans Control Syst Technol 11(5):765–772

Silva RN, Rato LM, Lemos JM (2003b) Time scaling internal predictive control of a solar plant. Control Eng Pract 11(12):1459–1467

Villadsen J, Michelsen ML (1978) Solution of differential equation models by polynomial approximations. Prentice-Hall, Englewood Cliffs

Wertz V, Bastin G, Haest M (1987) Identification of a glass tube drawing bench. In: Proceedings of 10th world congress on automatic control, vol 10. IFAC, Munich, pp 334–339

Chapter 3
Predictive Adaptive Control with Linear Models

Model Predictive control (MPC) is one of the most significant tools for controller design, not only in the process industry but in other emergent areas as well. Its importance stems from a number of features:

- A minimization of a cost function is involved and hence the controller can be designed such as to have a direct relationship with a performance index with economic significance;
- Since it depends on the prediction of the plant output over an enlarged time horizon, several desirable properties hold such as robustness with respect to plant input/output transport delay or the ability to stabilize plants that are simultaneously open-loop unstable and nonminimum phase;
- The possibility of incorporating constraints.

No attempt is made here to present a general overview of predictive control methods, and of the associated problems, a subject on which several books are available (Mosca 1995; Maciejowski 2002; Sánchez and Rodellar 1996). Instead, we concentrate on the aspects that have a more direct impact on adaptive control of distributed collector solar fields. In this chapter, we address controllers that rely on linear plant models and in Chap. 5 a class of nonlinear MPC directly related to DCSFs is described.

In a computer control framework, the basic idea of MPC consists in, at the beginning of each sampling interval, computing the value of the manipulated variable to apply to the plant by minimizing a cost defined along an horizon (called the prediction horizon) of future discrete time instants. In order to perform this minimization, the plant dynamics is described by predictive models that relate the samples of the manipulated variable with the samples of predicted values of the plant output along with the prediction horizon. A sequence of values of the manipulated variable results from this optimization, of which only the first is actually applied to the plant, the whole procedure being repeated at the begining of the next sampling interval. This procedure is known as "receding horizon" optimization (Kwon and Han 2005).

The origin of MPC can be traced back to the 1970s, or even early, and was strongly motivated by applications, specially processes in the fluid industry (Richalet et al. 1978; Cutler and Ramaker 1980).

J. M. Lemos et al., *Adaptive Control of Solar Energy Collector Systems*,
Advances in Industrial Control, DOI: 10.1007/978-3-319-06853-4_3,

In parallel to these developments, a different approach emerged that considers the plant to be represented by a discrete-time transfer function and subject to stochastic disturbances modeled by filtering a white noise sequence (that is to say, a sequence of independent identically random variables) by a linear transfer function with the same poles of the plant model. This is the so-called ARMAX model (Åstrom 1997).

For simplicity, consider the regulation problem that consists in keeping the plant output (considered as an increment with respect to a desired value) close to zero despite the effect of disturbances. The regulation control problem is formulated as the minimization of the plant output variance at the earliest sample affected by the current manipulated variable move (Åstrom 1973), that is to say, at discrete time k, the manipulated variable $u(k)$ is computed by minimizing the minimum variance cost J_{MV} given by

$$J_{\text{MV}}(u(k)) = \mathcal{E}\left[y^2(k+d)|\mathcal{I}^k\right], \tag{3.1}$$

where d is the plant pure delay and $\mathcal{E}[\cdot|I^k]$ stands for the mean conditioned on (the σ-algebra induced by) the information available up to time k, denoted $\mathcal{I}^k$. Adaptation is embedded by estimating the parameters of a d-steps ahead plant predictive model using recursive least squares (RLS) and using the estimates instead of the true plant parameters to compute the manipulated variable.

Since the minimum variance control law yields a closed loop that cancels the zeros of the open-loop plant (Åstrom 1997), if the plant is nonminimum phase there will be unstable internal modes. Even when the closed-loop system is stable, minimum variance control is too reactive and often results in excessive control action.

In order to overcome the limitations of the minimum variance control law, a new version was proposed (Clarke and Gawthrop 1975, 1979) in which the cost (3.1) is modified by the inclusion of a term that penalizes the manipulated variable, resulting in

$$J_{\text{detuned}}(u(k)) = \mathcal{E}\left[y^2(k+d) + \rho u^2(k)|\mathcal{I}^k\right] \tag{3.2}$$

with ρ a positive parameter (weight on the manipulated variable penalty).

Controller design based on the minimization of J_{detuned} (3.2) is an important modification because the extra term regularizes the control variable, providing the designer with a knob (the value of ρ) that changes the type of closed-loop response obtained.

The resulting control law can be used to stabilize open-loop stable nonminimum phase systems but is not able to stabilize a plant that is both open-loop unstable and nonminimum phase. In addition, one has to be known precisely the value of the plant pure delay d. MPC, that has a tight link with optimal control, solves these problems by considering a multistep cost function that is optimized in a receding horizon way using predictive models. Furthermore, MPC provides a natural setting for the inclusion of constraints in control design. These ideas are now explained with emphasis on the relation to distributed collector solar field application.

3.1 Receding Horizon Control

Since the aim is to work in a computer control framework, a discrete time formulation of models and control is considered. For that sake, let $t \in \mathbb{R}$ denote continuous time and let t_k be a sequence of continuous time instants corresponding to the sampling times, with k an integer that denotes the discrete time. The use of a sampling interval that varies along time will be explored in Chaps. 5 and 7. In this chapter, it is assumed that the sampling interval Δ is constant and hence

$$t_k = t_{k-1} + \Delta \tag{3.3}$$

Hereafter, except when stated otherwise, we only consider the variables at the sampling instants and, hence, only the discrete time index k is used.

In MPC, a multistep cost function such as

$$J_T = \mathcal{E}\left\{\sum_{j=T_1}^{T} \tilde{y}^2(k+j) + \rho \sum_{j=1}^{T_u} \hat{u}^2(k+j-1)|\mathcal{I}^k\right\} \tag{3.4}$$

is minimized at the beginning of each sampling interval k. The parameter T is called "prediction horizon" and T_u is called "control horizon." Parameter T_1 is included in order to avoid the unnecessary prediction of plant outputs in processes with pure delay bigger than 1. In many cases, we select $T_1 = 1$ to avoid making assumptions on the minimum value of the process delay.

Figure 3.1 shows the main variables used in MPC. These are the virtual manipulated variable $\hat{u}$, the plant output y and the reference to track r.

A virtual reference r^* is used to connect the present value of the measured plant output $y(k)$ with the desired reference value at the end of the prediction horizon, $r(k+T)$. A possible choice consists in obtaining $r^*(k+j)$ for $j = 1, \ldots, T$ by defining

$$r^*(k) = y(k) \tag{3.5}$$

and then computing recursively for $j = 1, \ldots T$ the following difference equation

$$r^*(k+j) = \gamma_r r^*(k+j-1) + (1-\gamma_r)r(k+T) \tag{3.6}$$

where $0 \le \gamma_r < 1$ is a parameter. This procedure is a command governor that smooths out the response to abrupt reference changes, preventing saturation and strong control moves. Smaller values of γ_r make the reference transitions faster, while values of γ_r close to 1 make the controller less aggressive. The tracking error is then defined for any k as

$$\tilde{y}(k) = y(k) - r^*(k) \tag{3.7}$$

It is assumed that one is at time k ("present time") and wants to compute the manipulated variable $u(k)$ to apply to the plant during that sampling interval. For

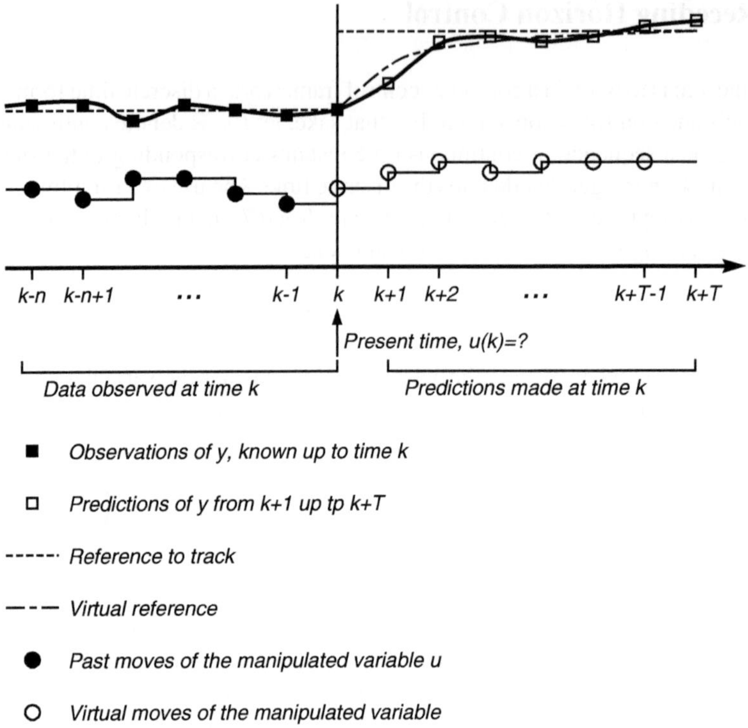

Fig. 3.1 The variables in receding horizon control

that sake, one minimizes J_T given by (3.4) with respect to the sequence of virtual future values of the manipulated variable, $\hat{u}(k), \ldots, \hat{u}(k+T-1)$, obtaining an optimal virtual control sequence, $\hat{u}^*(k), \ldots, \hat{u}^*(k+T-1)$, that minimizes (3.4). According to the receding horizon strategy (Kwon and Han 2005), the value of the manipulated variable at time k to actually apply to the plant is chosen to be the first value of this optimal sequence:

$$u(k) = \hat{u}^*(k). \tag{3.8}$$

The whole procedure is then repeated at the beginning of the next sampling interval.

Depending on the values of T, T_u and T_1, and on the existence and type of constraints, several control laws with quite different properties can be obtained. For instance, in the limit case where $T_1 = T = d$ and $T_u = 1$, the control law reduces to the one-step detuned minimum variance control yielded by (3.2). More interesting control laws (in the sense that they are able to stabilize a wider class of plants, being also more robust with respect to available plant knowledge) are obtained by extending both the prediction and control horizons.

Consider the case in which $T_u = T$ and there are no constraints. Furthermore, let the plant be represented by a reachable and detectable state-space model. In this situation, minimizing J_T is equivalent to a state feedback control law in which the gains are computed by iterating backwards T steps of an associated Riccati difference equation (Mosca 1995). Hence, if the horizon T is large enough, $u(k)$ will be given by a stabilizing state-feedback law whose gains are a tight approximation to the steady-state (infinite horizon) Linear Quadratic (LQ) gains (Mosca 1995). For the above choice of horizons, this fact provides a heuristic justification for the use of receding horizon control as a way to approximate steady-state LQ optimal control. Furthermore, this example also explains the limitations of minimum variance and detuned minimum variance control. Indeed, since minimizing (3.2) corresponds to just iterating one step of the Riccati equation, the resulting control may be quite different to the steady-state (infinite horizon) solution, a situation that happens, for instance, for plants that have zeros outside the unit circle.

When considering the above reasoning, an issue immediately arises: How large should be the horizon in order to ensure a stable closed-loop? From a practical point of view, one can adjust T by trial and error, for instance by making simulations. This procedure is however unsatisfactory from a theoretical point of view. Several answers have been given in the second half of the 1980s for linear systems, some of which inspired also solutions for stabilizing nonlinear systems. The solutions based on stability constraints, that lead to a higher computational load, are explained concisely in Sect. 3.5.2.

3.2 Linear Predictive Models

In order to minimize (3.4) the relation between the future plant manipulated input samples (that is to say, from time k up to time $k + T_u$) and the output samples (that is to say, from time $k + 1$ up to time $k + T$), is expressed by predictive models.

3.2.1 ARX Plant Model

In order to obtain predictive models, assume that the plant is represented by the difference equation

$$y(k+n) + a_1 y(k+n-1) + \cdots + a_n y(k) = b_0 u(k+m) + b_1 u(k+m-1) + b_m u(k) + e(k+n). \tag{3.9}$$

Here, u denotes the plant manipulated input, y denotes the plant output and e is a sequence of uncorrelated Gaussian variables of zero mean and variance σ_e^2 that models stochastic disturbances that affect the plant. In addition, n and m are integer numbers and $a_i, i = 1, \ldots, n$ and $b_i, i = 1, \ldots, m$ are plant dependent parameters. It is assumed that $n > m$ ir order to ensure that the process model is causal and has a pure delay of at least 1.

Model (3.8) is named in the literature as ARX (acronym of "Auto-Regressive with eXogenous input") (Åstrom 1997). It assumes that a sample of the plant input at a given time instant can be expressed as a linear combination of the previous n samples and n manipulated input samples delayed by $n-m$ samples. A closely related model is the ARMAX model, in which the disturbance is given by a linear combination of past samples of e, and not just by $e(k+n)$.

Defining the forward shift operator q as a mathematical operator that transforms a discrete time signal by advancing it by one sample, the ARX model can be written in compact form as

$$A(q)y(t) = B(q)u(t) + q^{n_a}e(t), \tag{3.10}$$

where $A(q)$ and $B(q)$ are polynomials in the forward shift operator given by

$$A(q) = q^n + \sum_{i=1}^{n} a_i q^{n-i} \tag{3.11}$$

and

$$B(q) = \sum_{i=0}^{m} b_i q^{m-i}. \tag{3.12}$$

Let now the backward shift operator q^{-1} be a mathematical operator that transforms a discrete time signal by delaying it by one sample and define the reciprocal polynomial of a polynomial $X(q)$ of degree n by

$$X^*(q^{-1}) \triangleq q^{-n}X(q) \tag{3.13}$$

With this definition, the ARX model (3.9) can also be written as:

$$A^*(q^{-1})y(k) = B^*(q^{-1})u(k-\delta_{io}) + e(k), \tag{3.14}$$

where δ_{io} is the plant pure input/output delay given by

$$\delta_{io} = n - m \tag{3.15}$$

and

$$A^*(q^{-1}) = 1 + \sum_{i=1}^{n} a_i q^{-i}, \tag{3.16}$$

$$B^*(q^{-1}) = \sum_{i=0}^{m} b_i q^{-i}. \tag{3.17}$$

It is assumed that $\delta_{io} = 1$. This does not imply any loss of generality since plants with longer delay can be considered by assuming some of the leading terms of $B^*(q^{-1})$ to be zero.

In addition, it is assumed that A and B are coprime polynomials, meaning that that they have no common roots. This assumption is required to ensure that the control design problem that is to be considered in association with the plant is well posed.

3.2.2 Positional Predictive Models

The theory of Diophantine equations (Åstrom 1997) guarantees that there exist unique polynomials

$$F_j^*(q^{-1}) = 1 + f_1 q^{-1} + \cdots + f_{j-1} q^{-j+1} \tag{3.18}$$

and

$$G_j^*(q^{-1}) = g_0^j + g_1^j q^{-1} + \cdots + g_{n-1}^j q^{-n+1} \tag{3.19}$$

that satisfy the identity

$$1 = F_j^*(q^{-1}) A^*(q^{-1}) + (q^{-j}) G_j^*(q^{-1}). \tag{3.20}$$

From a more elementary point of view, $F_j^*(q^{-1})$ is the quotient of order $j - 1$ of the long division of 1 by $A^*(q^{-1})$ and $G_j^*(q^{-1})$ is the corresponding remainder multiplied by q^j.

In order to obtain the j-steps ahead predictor for the plant output y, apply (that is to say, multiply in a formal sense, using the definition of q^{-1}) the operator polynomial $F_j^*(q^{-1})$ to the plant model (3.14) and use (3.18) to get

$$y(k+j) = G_j^*(q^{-1}) y(t) + F_j^*(q^{-1}) B^*(q^{-1}) u(k+j) + F_j^*(q^{-1}) e(t+j) \tag{3.21}$$

The optimal least squares predictor of $y(k+j)$ given observations up to time k is a quantity $\hat{y}(k+j|k)$ that minimizes the quadratic cost $J_P(\hat{y}(k+j|k))$ given by

$$J_P(\hat{y}(k+j|k)) = \mathcal{E}\left\{[y(k+j) - \hat{y}(k+j|k)]^2 | \mathcal{I}^k\right\} \tag{3.22}$$

Inserting (3.21) in (3.22), expanding the square and using the properties of the statistical average yields

$$\begin{aligned} J_P(\hat{y}(k+j|k)) = & \left(G_j^*(q^{-1})y(k) + F_j^*(q^{-1})B^*(q^{-1})u(k+j) - \hat{y}(k+j|k)\right)^2 \\ & + \mathcal{E}\left\{\left[F_j^*(q^{-1})e(k+j)\right]^2\right\} \\ & + 2\mathcal{E}\left\{\left[G_j^*(q^{-1})y(k) + F_j^*(q^{-1})B^*(q^{-1})u(k+j)\right]\right. \\ & \qquad \left. F_j^*(q^{-1})e(k+j)|\mathcal{I}^k\right\} \end{aligned} \tag{3.23}$$

Because $F_j^*(q^{-1})$ is a polynomial of order $j-1$, the term $F_j^*(q^{-1})e(k+j)$ is a linear combination of samples of e between times $k+1$ and $k+j$ and, hence, the cross term in (3.23) is the average of the product of two statistically independent terms or, equivalently, the product of their means. Since the mean of $F_j^*(q^{-1})$ is zero, it is concluded that the cross term is zero.

Furthermore, since e is a sequence of zero mean, independent, identically distributed variables, it follows that

$$\mathcal{E}\left\{\left[F_j^*(q^{-1})e(k+j)\right]^2\right\} = \left(1 + f_1^2 + \cdots + f_{j-1}^2\right)^2 \sigma_e^2 \tag{3.24}$$

Combining the above reasonings, it is concluded that

$$\begin{aligned} J_P(\hat{y}(k+j|k)) = & \left(G_j^*(q^{-1})y(k) + F_j^*(q^{-1})B^*(q^{-1})u(k+j) - \hat{y}(k+j|k)\right)^2 \\ & + \left(1 + f_1^2 + \cdots + f_{j-1}^2\right)^2 \sigma_e^2 \end{aligned} \tag{3.25}$$

The minimum of $J_P(\hat{y}(k+j|k))$ is therefore obtained by selecting the predictor as

$$\hat{y}(k+j|k)) = G_j^*(q^{-1})y(k) + F_j^*(q^{-1})B^*(q^{-1})u(k+j) \tag{3.26}$$

The corresponding least-squares prediction error variance is

$$\sigma_P^2 = \left(1 + f_1^2 + \cdots + f_{j-1}^2\right)^2 \sigma_e^2 \tag{3.27}$$

We now give the predictor (3.26) a form that can be more readily interpreted in terms of the response of a linear system and, at the same time is more prone to use in MPC. For that sake, define polynomials $\Xi_j^*(q^{-1})$ and $H_j^*(q^{-1})$ such as to decompose the product $F_j^*(q^{-1})B^*(q^{-1})$ as

$$F_j^*(q^{-1})B^*(q^{-1}) = H_j^*(q^{-1}) + q^{-j}\,\Xi_j^*(q^{-1}) \tag{3.28}$$

with

$$H_j^*(q^{-1}) = h_1 q^{-1} + \cdots + h_j q^{-j} \tag{3.29}$$

and

$$\Xi_j(q^{-1}) = \xi_0^j + \xi_1^j q^{-1} + \cdots + \xi_{m-1}^j q^{-m+1}. \tag{3.30}$$

Using (3.18) it is readily concluded that the coefficients of $H_j^*(q^{-1})$ are samples of the impulse response (also called Markov parameters) of the plant discrete transfer function A/B. With the above definitions, the predictor (3.26) is written

$$\hat{y}(k+j|k) = H_j^*(q^{-1})u(k+j) + G_j^*(q^{-1})y(k) + \Xi_j(q^{-1})u(k-1) \tag{3.31}$$

Define the vector of coefficients of the j-steps ahead predictive model

$$\pi_j \triangleq \begin{bmatrix} g_0^j & \cdots & g_{n-1}^j & \xi_0^j & \cdots & \xi_{m-1}^j \end{bmatrix}^T \tag{3.32}$$

and the *pseudostate*

$$s(k) \triangleq \begin{bmatrix} y(k) & \cdots & y(k-n+1) & u(k-1) & \cdots & u(k-m) \end{bmatrix}^T. \tag{3.33}$$

In the case of ARX plants, the pseudostate is actually a state, although not of minimum length. A state space realization associated to it is controllable but not observable. The case of ARMAX plants is discussed in Sect. 3.2.5.

With the above definitions, the j-steps ahead least squares predictor is written in vector notation

$$\hat{y}(k+j|k) = \sum_{i=1}^{j} h_i u(k+j-i) + \pi_j^T s(k). \tag{3.34}$$

Expression (3.34) has a straightforward interpretation. The predictor $\hat{y}(k+j|k)$ is the sum of two terms:

- The free response at time $k+j$ due to the decay of the initial condition of the state at time k, given by $\pi_j^T s(k)$ and
- The convolution of the future control moves, $u(k), \ldots, u(k+j-1)$, with the process impulse response samples, $h_1, \ldots, h_j$, given by $\sum_{i=1}^{j} h_i u(k+j-i)$.

Since it is assumed that one is at time k, the first free (that is to say, whose value is not yet decided) sample of control is $u(k)$. Furthermore, it is assumed that the nominal delay of the process is 1, which explains why the summation stops at $u(k+j-1)$ since $h_0 = 0$. This does not imply a loss of generality since, for plants with delays larger than 1 it simply happens that some of the leading terms of h_i vanish.

The actual value of $y(k+j)$ and its least squares predictor given observations up to time k are related by

$$y(k+j) = \hat{y}(k+j|k) + \varepsilon_j(k), \tag{3.35}$$

where $\varepsilon_j(k)$ is uncorrelated with $\hat{y}(t + j|t)$, meaning that it verifies the so called orthogonality condition

$$\mathcal{E}\left\{\hat{y}(k + j|k)\varepsilon_j(k)\right\} = 0. \tag{3.36}$$

Since MPC relies on sets of predictive models over a time horizon, it is useful to define a convenient notation to represent pencils of predictive models. Write the predictor (3.34) for $j = 1$ up to $j = T$:

$$\hat{y}(k + 1|t) = h_1 u(k) + \pi_1^T s(k) \tag{3.37}$$
$$\hat{y}(k + 2|t) = h_1 u(k + 1) + h_2 u(k) + \pi_2^T s(k) \tag{3.38}$$
$$\dots \tag{3.39}$$
$$\hat{y}(k + T|t) = h_1 u(k + T - 1) + \dots + h_T u(k) + \pi_T^T s(k) \tag{3.40}$$

and define the vectors

$$Y \triangleq \begin{bmatrix} y(k + 1) & \cdots & y(k + T) \end{bmatrix}^T, \tag{3.41}$$

$$\hat{Y} \triangleq \begin{bmatrix} \hat{y}(k + 1|k) & \cdots & \hat{y}(k + T|k) \end{bmatrix}^T, \tag{3.42}$$

$$U \triangleq \begin{bmatrix} u(k) & \cdots & u(k + T - 1) \end{bmatrix}^T, \tag{3.43}$$

$$E \triangleq \begin{bmatrix} \varepsilon_1(k) & \cdots & \varepsilon_T(k) \end{bmatrix}^T \tag{3.44}$$

and the matrices

$$H \triangleq \begin{bmatrix} h_1 & 0 & \cdots & 0 & 0 \\ h_2 & h_1 & \cdots & 0 & 0 \\ \cdots & \cdots & \cdots & \cdots & \cdots \\ h_T & h_{T-1} & \cdots & h_2 & h_1 \end{bmatrix} \tag{3.45}$$

and

$$\Pi \triangleq \begin{bmatrix} \pi_1| & \cdots & |\pi_T(k) \end{bmatrix}^T. \tag{3.46}$$

The set of predictive models is written in compact form as

$$\hat{Y} = HU + \Pi^T s(k) \tag{3.47}$$

and

$$Y = \hat{Y} + E. \tag{3.48}$$

Due to (3.36),

$$\mathcal{E}\left[E^T \hat{Y}\right] = 0. \tag{3.49}$$

Defining

$$Y_0 \triangleq \Pi^T s(k), \tag{3.50}$$

the set of predictive models (3.47) may also be written as

$$\hat{Y} = HU + Y_0. \tag{3.51}$$

3.2.3 Incremental Predictive Models

An alternative to the ARX model (3.14) consists in assuming that the plant is represented by

$$A^*(q^{-1})y(k) = B^*(q^{-1})u(k-1) + d_{\text{off}} + \eta(k), \tag{3.52}$$

where $A^*(q^{-1})$ and $B^*(q^{-1})$ are co-prime polynomials in the unit delay operator q^{-1}, parameterized as in (3.16, 3.17), d_{off} is an unknown constant offset and the stochastic disturbance η satisfies

$$\Delta(q^{-1})\eta(k) = e(k), \tag{3.53}$$

with the increment operator

$$\Delta(q^{-1}) \triangleq 1 - q^{-1} \tag{3.54}$$

and e is a white noise sequence as in model (3.14).

The assumption (3.53) made on the stochastic disturbance amounts to state that η is a random walk, and therefore that its power grows without bound when time increases, and may seem somewhat artificial. It is justified because the corresponding incremental version of the model involves only a white sequence. Actually, model (3.52) is the basis of the widespread GPC adaptive control algorithm (Clarke et al. 1987a).

Let $F_j^*(q^{-1})$ and $G_j^*(q^{-1})$ satisfy the following diophantine equation

$$1 - q^{-j} = \Delta(q^{-1})F_j^*(q^{-1})A_j^*(q^{-1}) + \Delta(q^{-1}) - q^{-j}G_j^*(q^{-1}). \tag{3.55}$$

It is remarked that a solution always exists because (Åstrom 1997) the maximum common divisor of $\Delta(q^{-1})A_j^*(q^{-1})$ and $\Delta(q^{-1}) - q^{-j}$ (that is equal to $\Delta(q^{-1})$) divides $1 - q^{-j}$.

Apply now $\Delta(q^{-1})F_j^*(q^{-1})$ to both sides of (3.52) and use (3.55) to write the j-steps ahead

$$\hat{y}(k+j|k)) = y(k) + G_j^*(q^{-1})\Delta(q^{-1})y(k) + F_j^*(q^{-1})B^*(q^{-1})\Delta(q^{-1})u(k+j). \tag{3.56}$$

Defining the incremental pseudostate by

$$s_\Delta(k) \triangleq \begin{bmatrix} y(k) & \Delta y(k) & \cdots & \Delta y(k-n+1) & \Delta u(k-1) & \cdots & \Delta u(k-m) \end{bmatrix}^T. \tag{3.57}$$

the incremental predictor is written

$$\hat{y}(k+j|k) = \sum_{i=1}^{j} h_i \Delta u(k+j-i) + \pi_j^T s_\Delta(k). \tag{3.58}$$

The pencil of predictors for $j = 1$ up to $j = T$ is written

$$\hat{Y} = HU_\Delta + Y_{\Delta,0} \tag{3.59}$$

where

$$U_\Delta \triangleq \begin{bmatrix} \Delta u(k) & \cdots & \Delta u(k+T-1) \end{bmatrix}^T, \tag{3.60}$$

and

$$Y_{\Delta,0} \triangleq \Pi^T s_\Delta(k). \tag{3.61}$$

3.2.4 MUSMAR Predictive Models

It is possible to consider predictive models that are alternative to the ones described in Sects. 3.2.2 and 3.2.3 and that are suitable for adaptive MPC algorithms that approximate the steady-state (infinite horizon) solution of the linear-quadratic gaussian (LQG) optimal control problem. This class of models, introduced in (Greco et al. 1984), is obtained by assuming that, along the control horizon, from $k+1$ up to $k+T-1$, the manipulated variable is constrained to be a constant feedback of the pseudostate. Using an innovations state-space model for the evolution of the pseudostate it is possible to eliminate the dependency of $y(k+j)$ on the samples of $u(k+i)$ for $i = 1$ and $i = j-1$. In order to use these predictors in MPC, it is also needed to predict future values of the control samples. The details are explained in Appendix C.

The above mentioned procedure yields the positional j-step ahead predictors for the output

$$\hat{y}(k+j|k) = \theta_j u(k) + \psi_j^T s(k) \tag{3.62}$$

and, for the future samples of the manipulated input

$$\hat{u}(k+j|k) = \mu_j u(k) + \phi_j^T s(k). \tag{3.63}$$

For ease of reference (3.62, 3.63), will hereafter be referred as "MUSMAR predictors" because they are associated with the predictive adaptive MUSMAR (acronym of Multistep, Multivariable, Adaptive Regulator) control algorithm, to be explained latter in Sect. 3.6. Here $\theta_j \in \mathbb{R}$, $\mu_j \in \mathbb{R}$, $\psi_j \in \mathbb{R}^{n_s}$ and $\phi_j \in \mathbb{R}^{n_s}$ are parameters and n_s is the dimension of the pseudo-state s.

In the "ideal" case in which the plant is represented by the ARX model (3.14), these parameters can be computed from the coefficients of the polynomials A and B of the ARX model, and from the feedback gain assumed to act between from $k+1$ up to $k+T-1$. In practice, however, the parameters of (3.62, 3.63) are estimated directly from plant data, using least squares and implying that

$$y(k+j) = \hat{y}(k+j|k) + \varepsilon_j^y(k) \tag{3.64}$$

and

$$u(k+j) = \hat{u}(k+j|k) + \varepsilon_j^u(k) \tag{3.65}$$

where $\varepsilon_j^y(k)$ is uncorrelated with $\hat{y}(t+j|t)$), and $\varepsilon_j^u(k)$ is uncorrelated with $\hat{u}(t+j|t)$) meaning that they verifies the orthogonality conditions

$$\mathcal{E}\left\{\hat{y}(k+j|k)\varepsilon_j^y(k)\right\} = 0 \tag{3.66}$$

and

$$\mathcal{E}\left\{\hat{u}(k+j|k)\varepsilon_j^u(k)\right\} = 0. \tag{3.67}$$

Using a common nomenclature, we say that variables such as $\hat{y}(k+j|k)$ and $y(k+j)$ that differ between themselves by a term that is uncorrelated to the regressor in $\hat{y}(k+j|k)$ are "equal in least-squares sense" and use the notation

$$y(k+j) \approx \hat{y}(k+j|k). \tag{3.68}$$

In addition to simplify the algorithm, the above procedure has the major advantage that the variables that enter the pseudo-state s can be chosen in a flexible way, without requiring the computation of predictive models with unknown dynamics. Thereby the designer is allowed the possibility of imposing the structure of the controller, by selecting n and m, by including feedforward terms from accessible disturbances, or state variables that are measured. It should be clear, however, that some choices may lead to controllers that are not able to stabilize the plant whatever the value of their gains.

The drawback of the predictive model (3.62, 3.63) is that, apart from $j = 1$, their parameters depend not only on the plant, but on the feedback applied as well. As a consequence if, for instance, the manipulated variable saturates for long periods, (3.62, 3.63), with their parameters update by a recursive estimator, will loose their capability to correctly predict y and u.

Incremental models in the case of MUSMAR predictors may also be obtained, being of the form

$$\hat{u}(k + j|k) = \mu_j \Delta u(k) + \phi_j^T s_\Delta(k) \tag{3.69}$$

for the output predictors, with s_Δ given by (3.57), and

$$\hat{\Delta u}(k + j|k) = \mu_j \Delta u(k) + \phi_j^T s_\Delta(k) \tag{3.70}$$

for the increments of the manipulated variable.

3.2.5 *The Case of Colored Noise*

For ARMAX plants s is no longer a state and the least squares predictors of y may no longer be expressed as a linear combination of a finite number of past samples of y and u. However, it is possible to show (Mosca and Zappa 1989) that, in closed-loop and under a constant feedback gain, the plant still admits a parametrization such that the prediction errors still verify (3.66, 3.67) for MUSMAR predictors. Hence, even in the case of the ARMAX plants, the parameters of the MUSMAR predictive models can be estimated without bias with the standard least squares estimation methods. Thus, there is no need to use estimation methods that require approximations, such as extended least squares or maximum likelihood.

3.3 Predictive Model Identification

The predictive models in the preceding sections can be written in the form of the linear regression model

$$z(i) = \vartheta^T \varphi(i) + v(i) \tag{3.71}$$

where $z(i) \in \mathbb{R}$ and $\varphi(i) \in \mathbb{R}^{n_p}$ are variables that can be measured for each index i, ϑ is a vector of parameters to estimate, n_p is the number of parameters and $v(i)$ is a residue that accounts for mismatches between the linear model (3.71) and the actually observed data.

The problem to consider consists in estimating the vector of parameters ϑ from a set of data $z(i)$, $\varphi(i)$ for $i = 1, \ldots, k$ with k the total number of observations.

For instance, to estimate the j-steps ahead predictor in (3.64), we select $z(k) = y(k + j)$, $\varphi(k) = [u(k)\ s^T(k)]^T$ and $\vartheta = [\theta\ \phi^T]^T$. Similar choices are made to the other predictive models.

3.3.1 Recursive Least Squares Estimation

The least squares estimate of the parameter vector ϑ in the linear regression model (3.71), given k observation points, is such that it minimizes the cost functional $J : \mathbb{R}^{n_p} \to \mathbb{R}$ defined by

$$J_{\mathrm{LS}}(\vartheta) = \frac{1}{2}\sum_{i=1}^{k} \lambda^{k-i}[z(i) - \varphi^T(i)\vartheta]^2. \tag{3.72}$$

Here, λ is number selected between 0 and 1 called the "forgetting factor". If $\lambda = 1$ all the observation points are weighted equally in J_{LS}. If $\lambda < 1$, the more recent (meaning that its time index is closer to k) is the observation point, the greater is the weight associated to it. This means that observations made in the past have a smaller influence on J_{LS} than more recent ones. Since the weight varies exponential in time, the resulting algorithm is called "least squares with exponential forgetting".

It is shown in Appendix B that the least squares estimate $\hat{\vartheta}(k)$ of ϑ given observations up to time k is computed recursively from the estimate given observations up to time $k-1$ by

$$\hat{\vartheta}(k) = \hat{\vartheta}(k-1) + K(k)[y(k) - \varphi^T(k)\hat{\vartheta}(k-1)], \tag{3.73}$$

in which $K(k)$ is called the Kalman gain, being given by

$$K(k) = P(k)\varphi(k) \tag{3.74}$$

where the covariance matrix P satisfies the Riccati equation

$$P(k) = \left\{P(k-1) - \frac{P(k-1)\varphi(k)\varphi^T(k)P(k-1)}{\lambda + \varphi^T(k)P(k-1)\varphi(k)}\right\}\frac{1}{\lambda}. \tag{3.75}$$

Equations (3.73, 3.74, 3.75) define the RLS estimation algorithm with exponential forgetting. If no a priori information is available about the true value of ϑ, its estimate is initialized by selecting an arbitrary vector and $P(0)$ selected such as to reflect this lack of knowledge, for instance $P(0) = 1000I$, where I is the identity matrix. Since this choice may induce strong initial transients, this issue is discussed in much more detail in Sect. 3.9.

The above exponential forgetting algorithm defined by (3.73, 3.74, 3.75) implies that old data is discarded no matter the relation it has with the parameters to estimate. If some parameters have a degree of identifiability which is higher than others, forgetting data uniformly may led to some entries of the covariance matrix to grow unbound, leading to strong variation of the estimates. The application of Bayesian methods solved this problem by introducing algorithms with "directional forgetting" (Kulhavý 1987), in which data is discarded only along the directions of space that are aligned with the directions of incoming data. One possible algorithm for RLS with directional forgetting reads as follows

$$K(k) = \frac{P(k-1)\varphi(k)}{1 + \varphi'(k)P(k-1)\varphi(k)[1 - \beta_{\mathrm{df}}(k)]}, \tag{3.76}$$

$$P(k) = [I - K(k)\varphi'(k)(1 - \beta_{\mathrm{df}}(k))]P(k) \tag{3.77}$$

and

$$\hat{\vartheta}(k) = \hat{\vartheta}(k-1) + K(k)[y(k) - \hat{\vartheta}(k)'\varphi(k)]. \tag{3.78}$$

The variable $\beta_{\mathrm{df}}(k)$ denotes the quantity of information discarded in each iteration, being given according to a directional forgetting (Kulhavý 1987) scheme by

$$\beta_{\mathrm{df}}(k) = 1 - \lambda + \frac{1-\lambda}{\varphi'(k)P(k-1)\varphi(k)}, \tag{3.79}$$

where λ is a constant to be chosen between 0 (complete forgetting) and 1 (no forgetting) that determines the rate of forgetting in the direction of incoming information.

The equations that propagate the covariance matrix P in RLS either with exponential forgetting (3.75) or with directional forgetting (3.77) are ill-conditioned from a numerical point of view since they imply the subtraction of two quantities whose difference is much smaller than their absolute value. This, in practice, a factorized version is used (Bierman 1977) in which the matrix P is factorized as $P = UDU^T$ where U is an upper triangular matrix with the diagonal elements equal to 1 and D is a diagonal matrix. Instead of P, matrices U and D are propagated in time. The major advantage is that the resulting equivalent P is always positive definite and symmetric, no matter the numerical errors induced by roundoff.

3.3.2 Predictor Extrapolation and Redundant Estimation

In order to obtain the coefficients of the predictors (3.31) one possibility (Clarke et al. 1987a) is to estimate the parameters of the one-step ahead predictor by RLS and then to compute the parameters of the other predictors by extrapolation using (3.20). This can be done with the following algorithm:

3.3.2.1 Predictor Extrapolation

1. Using RLS, estimate the parameters in the one-step ahead predictor

$$\hat{y}(k+1|k) = h_1 u(k) + \pi_1^T s(k).$$

2. Let

$$a_i = -\pi_1(i) \text{ for } i = 0, \ldots, n$$

$$b_0 = h_1 \text{ and } b_i = \pi_1(i+n) \text{ for } i = 0, \ldots, m$$

3. Let $f_0 = 1$, and $g_i^0 = -a_{i+1}$, for $i = 0, \ldots, n-1$.
4. For $j = 1$ up to T execute recursively

$$f_j = g_0^{j-1}$$

$$g_i^j = g_{i+1}^{j-1} - a_{i+1} f_j \text{ for } i = 0, \ldots, n-1$$

$$g_{n-1}^j = -a_n f_j$$

$$F_j^*(q^{-1}) = 1 + \sum_{i=1}^{j} f_j q^{-i}$$

$$B^*(q^{-1}) = \sum_{i=0}^{m} b_i q^{-i}$$

Using the nomenclature in (3.28–3.30), set

$$\pi_{j+1}(i) = g_{i-1}^j \text{ for } i = 0, \ldots, n \text{ and}$$

$$\pi_{j+1}(n+i) = \xi_i^j \text{ for } i = 0, \ldots, m.$$

□

Above, $\pi_j(i)$ denotes the entry i of vector π_j. A formally equal procedure can be used to compute the parameters for incremental predictors. If the exact values of the parameters of the one-step ahead predictor are known, the above algorithm provides the correct value of the parameters of predictors over longer horizons. This is not the case if the model order is not correct or if there are plant nonlinearities. Furthermore, the plant pure delay must be known exactly in order to use the algorithm.

If, as it happens in adaptive control, the parameters of the one-step ahead predictor are estimated with RLS there will be an estimation error that propagates and is amplified when computing predictors over larger horizons. Furthermore, RLS is only unbiased estimator for ARX plants. For ARMAX plants RLS will yield biased estimates causing a significant deviation of the estimates from their correct values that is not compensated when combining the estimator with a MPC control law (opposite to what happens when considering the classical RLS plus minimum variance self tuner briefly discussed in the introduction of this chapter).

Another possibility that circumvents these problems is to separately estimate the parameters for all parameters. For the MUSMAR parameters (3.62) does not entail a big increase of the computational load since the regressor is the same for all the predictive models, and is given by $\varphi(k) = [u(k)\ s^T(k)]^T$. Therefore, only one covariance matrix must be propagated in time.

In the ideal situation in which the plant is linear with a known order, separate estimation of the predictors is redundant and the parameters of the predictive models are not identifiable. Unidentifiability of the parameters is a drawback that can be solved by using RLS with directional forgetting. The advantage is that, in practical situations, improves predictive performance since extrapolating the one-stepa ahead predictor yields a bigger error than direct estimation. Furthermore, as discuss below in Sect. 3.6.2, separate predictor identification with RLS leads to an adaptive control algorithm that, at least locally, adjusts the controller gains such as to minimize the underlying cost and progressing a Newton direction (Mosca et al. 1989).

3.4 The GPC Algorithm

Assuming that there are not inequality constraints and that the plant is described by a linear model, the control law that minimizes the quadratic cost functional (3.4) can be computed explicitly using knowledge of the predictive model parameters. Depending on the different type of predictive models used, different control laws will be issued.

If the plant is described by the pencil of incremental predictive models (3.51), the Generalized Predictive Control (GPC) law will be issued (Clarke et al. 1987a, b). To deduce this control law, write (3.4) in the form

$$J_T = \mathcal{E}\left\{\left(Y - Y^*\right)^T \left(Y - Y^*\right) + \rho U_\Delta^T U_\Delta | \mathcal{I}^k\right\}, \tag{3.80}$$

where Y is a vector of future plant output samples, as in (3.41), Y^* is a vector of corresponding future virtual references and U_Δ is a vector as in (3.60), but with the control increment samples replaced by virtual control increment samples. Inserting the vector predictive model (3.59) in (3.80) and using the fact that the residue vector E is uncorrelated with the predictor, yields

$$\begin{aligned} J_T = {} & U_\Delta^T M U_\Delta + 2\left(Y_{\Delta,0} - Y^*\right)^T H U_\Delta + \left(Y_{\Delta,0} - Y^*\right)^T \left(Y_{\Delta,0} - Y^*\right) \\ & + \mathcal{E}\left\{E^T E\right\}, \end{aligned} \tag{3.81}$$

where

$$M \triangleq H^T H + \rho I, \tag{3.82}$$

with H the matrix of plant Markov parameters given by (3.45).

In order to find the value of U_Δ that minimizes J_T consider a generic quadratic function of U_Δ with minimum at U_Δ^*, written as

$$J_{\text{gen}} = \left(U_\Delta - U_\Delta^*\right)^T M \left(U_\Delta - U_\Delta^*\right) \tag{3.83}$$

and expand it to get

$$J_{\text{gen}} = U_\Delta^T M U_\Delta - 2U_\Delta^{*T} M U_\Delta + U_\Delta^{*T} U_\Delta^*. \tag{3.84}$$

Comparing the second term of (3.84) with the second term of (3.81), yields the optimal sequence of virtual control

$$U_\Delta^* = -M^{-1} H^T \left(Y_{\Delta,0} - Y^*\right). \tag{3.85}$$

According to the receding horizon strategy explained in Sect. 3.1, only the first element of this vector of virtual manipulated variables is actually applied to the plant. This is $\Delta u(k)$, the GPC control law is written as

$$u(k) = u(k-1) - \bar{e}_1^T M^{-1} H^T \left(Y_{\Delta,0} - Y^*\right), \tag{3.86}$$

where $\bar{e}_1^T = \begin{bmatrix} 1 & 0 & \cdots & 0 \end{bmatrix}$.

By recalling (3.61), the GPC control law can also be written as a feedback from the pseudostate plus a feedforward term

$$u(k) = u(k-1) - F^T s_\Delta(k) + u_{ff}(k), \tag{3.87}$$

where the vector of feedback gains, F, is given by

$$F = \Pi H M^{-1} \bar{e}_1 \tag{3.88}$$

and the feedforward term is

$$u_{\text{ff}}(k) = \bar{e}_1^T M^{-1} H^T Y^*. \tag{3.89}$$

Figure 3.2 summarizes the structure of the GPC control algorithm applied to an ARX plant. At the beginning of each sampling interval the estimates of the parameters of the one step-ahead plant output predictor are updated using the RLS algorithm, either with exponential forgetting (Eqs. (3.73–3.75)) or with directional forgetting (Eqs. (3.76–3.79). From these estimates, the parameters of multiple steps ahead predictors are obtained by extrapolation using the algorithm os Sect. 3.3.2. From the estimates of the predictor parameters and the virtual reference compute the manipulated variable using (3.86) or, equivalently (3.87). If the plant delay is not 1, make T_1 equal to the plant delay, estimate by RLS the parameters of the T_1-steps ahead predictor and obtain the remaining predictors by extrapolation.

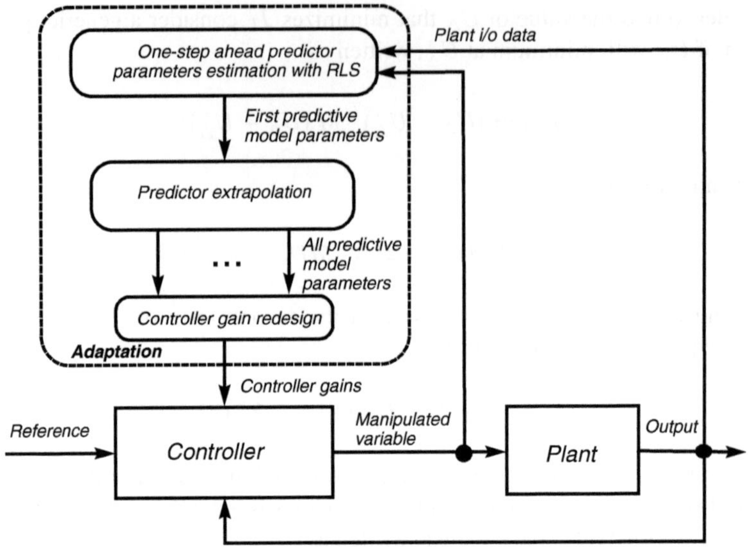

Fig. 3.2 Structure of the GPC adaptive control algorithm

3.5 Constraints

There are two types of constraints:

- Operational constraints;
- Stability constraints.

3.5.1 *Operational Constraints*

Operational constraints are constraints added to the problem of minimizing the cost in order to comply with operational specifications. For instance in a DCSF, the fluid flow must be kept between a minimum and a maximum value, and the temperature may not exceed a certain safety value. Thus, in general, the problem of computing the manipulated variable is formulated as

$$\min_{U_\Delta} J_T(U_\Delta) \tag{3.90}$$

subject to

$$A_u U_\Delta - b_u \leq 0 \tag{3.91}$$

and

$$a_y Y - b_y \leq 0 \tag{3.92}$$

where A_u, b_u, A_y and b_y are matrices and the inequalities are taken for each entry of the vectors. In order to solve this problem, one must resort to numerical methods. The references (Adetola et al. 2009; Kim and Sugie 2008; Kim et al. 2008; Maniar et al. 1997; Nikolakopoulos et al. 2006; Tanaskovic et al. 2013; Veres and Norton 1993) provide several adaptive predictive controllers that incorporate hard constraints such as these ones.

3.5.2 Stability Constraints

A major concern is to ensure the stability of the closed-loop resulting from the control algorithm being used. Without any constraints, the control yielded by the minimization of a cost defined over a finite horizon does not necessarily ensure stability.

When the control horizon T_u is equal to the prediction horizon T, GPC applied to the regulation problem (meaning that the reference to track is zero) yields feedback gains that are equal to the ones obtained by iterating backwards T steps of the Riccati equation associated to an equivalent LQ control optimization problem (Clarke et al. 1987a; Mosca 1995). The gains that result from the solution of the LQ problem for an infinite horizon stabilize the closed-loop but, for short values of the prediction horizon, the resulting gains may be quite different from the steady-state ones. In practice, it is expected that, by enlarging the horizon T, a better approximation of the steady-state LQ gains is obtained and GPC yields stabilizing gains.

The desire to provide conditions on the value of T to obtain stabilizing gains lead to a number of modifications of the basic MPC algorithms (Mosca 1995; Bitmead et al. 1990). The constraint that the plant state vanishes in the end of the optimization horizon solves this problem and gives rise to algorithms like SIORHC (for "Stable Input–output Receding Horizon Controller") (Mosca 1995). It is remarked that in the linear case imposing this constraint does not cause any numerical difficulties since the solution may still be obtained in closed-form.

3.6 The MUSMAR Algorithm

The Multistep, Multivariable Adaptive Regulator (MUSMAR) was designed (Greco et al. 1984) so as to approximate the Kleinman's iterations to solve numerically the steady-state Riccati equation associated to the minimization of J_T when $T \to \infty$.

3.6.1 The Basic Algorithm

The MUSMAR algorithm relies on the predictive models (3.62) described in Sect. 3.2.4, that are used to minimize the cost (3.4) in a receding horizon sense, and reads as follows for positional models:

3.6.1.1 MUSMAR Positional Algorithm

At the beginning of each sampling interval k (discrete time), recursively perform the following steps:
1. Sample plant output, $y(k)$ and use 3.7 to compute the tracking error $\tilde{y}$, with respect to the virtual set-point r^*.
2. Using RLS with directional forgetting, update the estimates of the parameters θ_j, ψ_j, μ_{j-1} and ϕ_{j-1} in the following set of predictive models

$$\tilde{y}(k+j) \approx \theta_j u(k) + \psi_j' s(k) \tag{3.93}$$

$$u(k+j-1) \approx \mu_{j-1} u(k) + \phi_{j-1}' s(k) \tag{3.94}$$

for

$$j = 1, \ldots, T.$$

Here, $\approx$ denotes equality in least squares sense and $s(k)$ is given by

$$\begin{aligned} s(k) = [\tilde{y}(k) \; \cdots \; \tilde{y}(k-n+1) \; u(k-1) \; \cdots \; u(k-m) \\ d(k) \; \cdots \; d(k-n_{w1})]^T \end{aligned} \tag{3.95}$$

where the d are samples of the feedforward signal incoming from accessible disturbances. Since, at time k, $\tilde{y}(k+j)$ and $u(k+j)$ are not available for $j \geq 1$, for the purpose of estimating the parameters, the variables in (3.93) are delayed in block of T samples. The RLS estimation equations are thus,

$$K(k) = \frac{P(k-1)\varphi(k-T)}{1+\varphi'(k-T)P(k-1)\varphi(k-T)[1-\beta(k)]} \tag{3.96}$$

and

$$P(k) = [I - K(k)\varphi'(k-T)(1-\beta(k))]P(k-1) \tag{3.97}$$

where, for $j = 1, \ldots, T$,

$$\hat{\Theta}_j(k) = \hat{\Theta}_j(k-1) + K(k)[y(k-T+j) - \hat{\Theta}_j(k-T)'\varphi(k-T)] \tag{3.98}$$

and for $j = 1, \ldots, T-1$

$$\hat{\Omega}_j(k) = \hat{\Omega}_j(k-1) + K(k)[u(k-T+j) - \hat{\Omega}_j(k-T)'\varphi(k-T)]. \tag{3.99}$$

In these equations, $\hat{\Theta}_j$ represents the estimate of the parameter vector of the output predictors, given at each discrete time k and for each predictor j by

$$\hat{\Theta}_j = [\theta_j \ \psi_j']' \tag{3.100}$$

and $\varphi(k - T)$ represents the regressor, common to all predictors, given by

$$\varphi(t - T) = [u(t - T) \, s'(t - T)]'. \tag{3.101}$$

Similarly, $\hat{\Omega}_j$ represents the estimate of the parameter vector of the input predictors, given at each discrete time and for each predictor j by

$$\hat{\Omega}_j = [\mu_j \ \phi_j']' \tag{3.102}$$

Since the regressor $\varphi(k - T)$ is common to all the predictive models, the Kalman gain update (3.96) and the covariance matrix update (3.97) are also common to all the predictors and need to be performed only once per time iteration. This computational organization greatly reduces the computational load.
3. Apply to the plant the control given by

$$u(t) = F's(t) + \eta(t) \tag{3.103}$$

where η is a white dither noise of small amplitude and F is the vector of controller gains, computed from the estimates of the predictive models by

$$F = -\frac{1}{\alpha}\left(\sum_{j=1}^{T} \theta_j \psi_j + \rho \sum_{j=1}^{T-1} \mu_j \phi_j\right), \tag{3.104}$$

with the normalization factor $\alpha > 0$ given by

$$\alpha = \sum_{j=1}^{T} \theta_j^2 + \rho\left(1 + \sum_{j=1}^{T-1} \mu_j^2\right). \tag{3.105}$$

□

Figure 3.3 shows the structure of the MUSMAR controller. The core of the algorithm is a linear controller that computes the manipulated variable through a feedback of the pseudo-state. The gains of this controller are adapted by an adaptation mechanism that consists of two major blocks. One of the blocks estimates the parameters of a set of predictive models. Using these estimates, according to a certainty equivalence strategy, the other block updates the gains of the linear controller.

An incremental version of MUSMAR can be obtained as well, in a similar way as for GPC, by using the MUSMAR incremental predictors (3.69, 3.70).

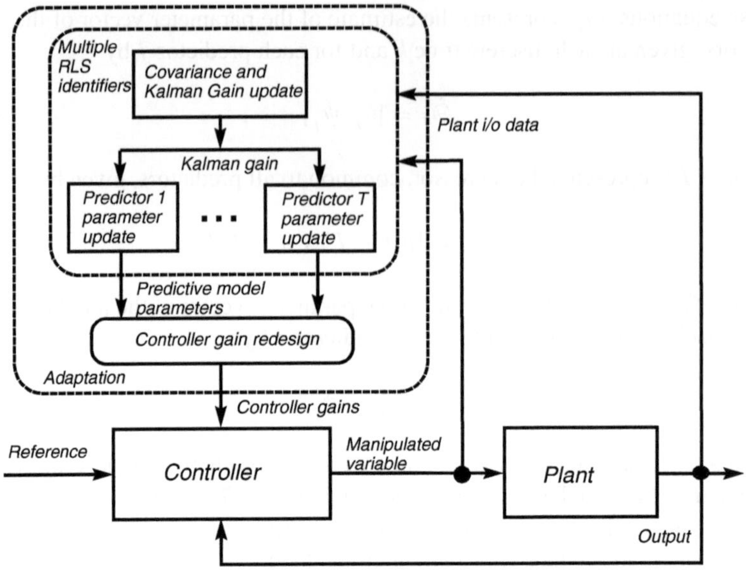

Fig. 3.3 Structure of the MUSMAR controller

3.6.2 MUSMAR as a Cost Optimizer

The convergence properties of the adaptive algorithm applied to each of the local controllers are summarized in the following propositions whose proofs are shown in the Appendix F.

Proposition 1 Mosca et al. (1989) *Let F_{k-1} be the controller gains computed by the MUSMAR algorithm. These gains are supposed to be applied to the controller for a number of past iterations long enough such that the estimator has converged and let the vector of update gains F_k be computed from* (3.104) *in such a way that the predictor parameters are replaced by their least-squares estimates at equilibrium. Then,*

$$F_k = F_{k-1} - \frac{1}{2\alpha(F_{k-1})} R_s^{-1}(F_{k-1}) \nabla_T J_T(F_{k-1}), \tag{3.106}$$

where

$$\alpha(F_{k-1}) \triangleq \sum_{i=1}^{T} \theta_i^2(F_{k-1}) + \rho \left(1 + \sum_{i=1}^{T-1} \mu_i^2(F_{k-1}) \right), \tag{3.107}$$

and

$$R_s(F_{k-1}) \triangleq \lim_{t\to\infty} E[s(t)s'(t)] \tag{3.108}$$

and $\nabla_T(F_{k-1})$ *is the gradient of the receding horizon cost* (3.4), *computed for* $F = F_{k-1}$. □

This proposition states that, if the prediction horizon T is large enough, each of the local adaptive controllers adjust the gains in a "Newton direction" in the process of minimizing the underlying steady-state (infinite horizon) quadratic cost

$$J_\infty = \lim_{k\to\infty} \mathcal{E}\left\{y^2(k) + \rho u^2(k-1)\right\} \tag{3.109}$$

In general, nothing can be proved about the convergence properties of the MUSMAR algorithm. However, if T is large enough, $\nabla_T J_i(F)$ is a tight approximation to the gradient of the steady state cost (3.109), $\nabla J_\infty(F)$, and the controller gains are then updated by

$$F_k = F_{k-1} - \frac{1}{2\alpha(F_{k-1})} R_s^{-1}(F_{k-1}) \nabla J_\infty(F_{k-1}) \tag{3.110}$$

In this limit case, the following proposition holds:

Proposition 2 Mosca et al. (1989) *Consider the regulation problem (zero reference). For the controller with the controller gains updated by* (3.104, 3.105), *the only possible convergence gains are local minima of the steady-state underlying cost* J_∞, *constrained to the local controller structure imposed by the choice of the pseudo-state* s. □

3.6.3 Constraints in MUSMAR

Constraints in the instantaneous value of either the input or the output cannot be incorporated in MUSMAR because, in such case, the assumption of a constant gain underlying MUSMAR predictors would not be valid. However, it is possible to impose constraints in the average vale of the manipulated variable.

Consider the problem of minimizing the steady state value of the plant output

$$\min \lim_{k\to\infty} \mathcal{E}\left[y^2(k)\right] \tag{3.111}$$

subject to the constraint on the input power

$$\mathcal{E}\left[u^2(k)\right] \le c^2, \tag{3.112}$$

where c is a constant.

The above problem can be solved in an approximate way using the MUSMAR algorithm and updating the control weight ρ along discrete time k using

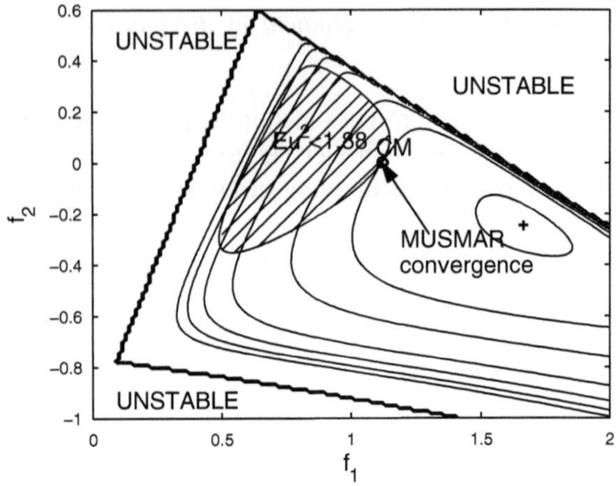

Fig. 3.4 Simulation example of MUSMAR with constraints. Convergence point of the controller gains superimposed on the level contour curves of the steady state cost

$$\rho(k) = \rho(k-1) + \varepsilon_c \rho(k-1) \left(u^2(k-1) - c^2 \right), \tag{3.113}$$

where ε_c is a small parameter.

As a simulation example of the use of the MUSMAR algorithm with constraints, consider the ARMAX plant defined by the difference equation

$$y(k) = -y(k-1) - 0.4y(k-2) + u(k-1) + e(k) - 0.7e(k-1). \tag{3.114}$$

Although this plant is of second order, the selected controller structure matchs a first order plant, and is defined by

$$u(k) = f_1 y(k) + f_2 u(k-1). \tag{3.115}$$

The controller gains f_1 and f_2 are to be adaptively designed such as to minimize the steady state variance of the output (3.111), subject to the constraint (3.112) with $c^2 = 1.88$. Figure 3.4 shows the contour lines of the cost (3.111). This function has a minimum at the point indicated by the sign + in Fig. 3.4. The shadowed region indicates the range of values of f_1 and f_2 that satisfy the constraint. The constrained minimum (CM) is indicated by the sign 'o'.

Figure 3.5 depicts the time evolution of the controller gains computed by MUSMAR with the weight ρ updated by (3.113). The thin straight lines indicate the optimum constrained values of $f_1 = 1.666$ and $f_2 - 0.247$ (that are unknown to the algorithm and have been included for the sake of comparison). The average of the values of the gains yielded by MUSMAR after convergence is shown by the sign $*$ in Fig. 3.4. As can be seen, MUSMAR tightly approximates the constrained minimum.

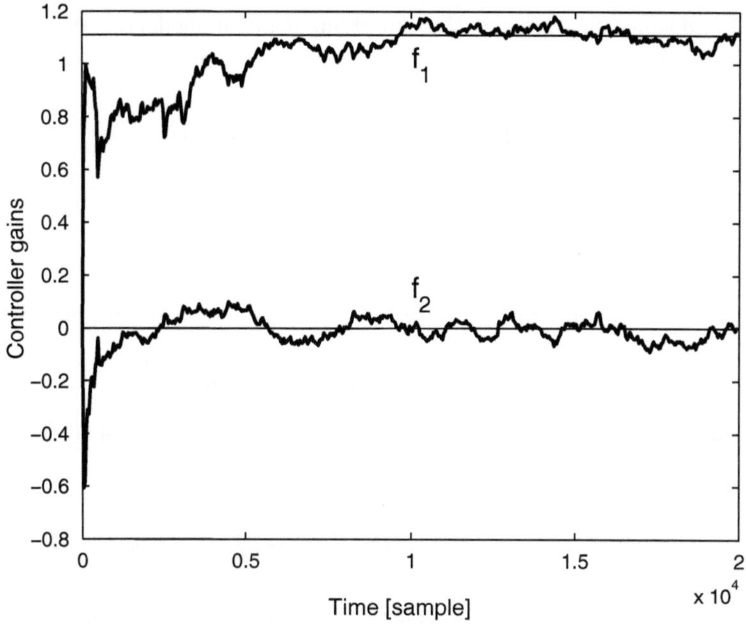

Fig. 3.5 Simulation example of MUSMAR with constraints. Evolution of the controller gains. The horizon *straight lines* indicate the optimal gains for comparison

3.6.4 Dynamic Cost

The design of controllers based on a quadratic cost such as (3.109) or its receding horizon approximation (3.4) can be improved by the inclusion of dynamic weights. This modification allows both to shape the response obtained for the closed loop system and to enhance the controller robustness with respect to unmodeled plant dynamics. In order to understand how this can be done, let the plant be described by the input–output model

$$A(q)y(k) = B(q)u(k) + d(k), \tag{3.116}$$

in which A and B are polynomials in the ward shift operator q, k denotes discrete time, u is the manipulated variable, y is the process output and d is a disturbance.

Let $H(q)$ be a polynomial such that $H(1) = 1$ and define the filtered variables

$$y_H(k) = \frac{H(q)}{q^{\partial H}} y(k) \tag{3.117}$$

and

$$u_H(k) = \frac{H(q)}{q^{\partial H}} u(k), \tag{3.118}$$

where ∂H denotes the degree of H. Multiplying the plant model (3.116) by $\frac{H(q)}{q^{\partial H}}$, it is recognized that the filtered variables are related by the model

$$A(q)y_H(k) = B(q)u_H(k) + \frac{H(q)}{q^{\partial H}}d(k). \tag{3.119}$$

Consider now the causal linear controller

$$r(q)u_H(k) = -S(q)y_H(k) + T(q)r(k), \tag{3.120}$$

in which r is the reference to track and R, S and T are polynomials in q that minimize

$$J_H = \lim_{k\to\infty} \mathcal{E}\left\{(H(q)y(k) - r(k))^2 + \rho(H(q)u(k)^2\right\}. \tag{3.121}$$

For generic values of R, S and T, the closed-loop model is obtained by eliminating u_H in (3.119) and (3.120), and is given by

$$y(k) = \frac{BT}{(AR+BS)H}r(k) + \frac{R}{AR+BS}\frac{1}{q^{\partial H}}d(k). \tag{3.122}$$

Let $d(k) = C(q)e(k)$ with e a white noise sequence (ARMAX model). In this case, if R, S and T minimize J_H, then they satisfy (Mosca 1995)

$$T(q) = C(q)H(q) \tag{3.123}$$

and the linear diophantine equation

$$A(q)R(q) + B(q)S(q) = Q(q)C(q)H(q), \tag{3.124}$$

where $Q(q)$ is a monic, hurwitz polynomial obtained by solving the spectral factorization problem

$$\gamma Q(z)Q(z^{-1}) = \rho A(z)A(z^{-1}) + B(z)B(z^{-1}), \tag{3.125}$$

where γ is a normalizing constant. Therefore, for the the optimal controller, using (3.123) and (3.124) in (3.122), the closed-loop model becomes

$$y(k) = \frac{B}{Q}\cdot\frac{1}{H}r(k) + \frac{R}{QC}\cdot\frac{q^{\partial H}}{H(q)}d(k). \tag{3.126}$$

From (3.126) it is concluded that the control law that results from the optimization of (3.121) causes the plant output to track the reference filtered by $1/H(q)$. Furthermore, the effect of the disturbance is also filtered by $1/H(q)$. Since H is a polynomial, $1/H(q)$ is a low pass filter and the high frequency components of the disturbance are therefore washed out.

In order to use a MUSMAR type approximation of the steady-state quadratic cost with dynamic weights (3.121), one should use the basic MUSMAR algorithm described in Sect. 3.6.1 with y and u replaced by its filtered versions. From the resulting u_H compute then the control to apply to the plant by

$$u(k) = \frac{q^{\partial H}}{H(q)} u_H(k). \tag{3.127}$$

As shown below, the inclusion of dynamic weights yields a significant improvement when controlling a DCSF because of the high frequency dynamics that is inevitably neglected with a finite dimensional model.

3.7 Using MUSMAR in Practical Examples

Before considering the application of MUSMAR to DCSFs, three other examples are considered that the following plants described in Chap. 2:

- Air heating fan (Sect. 2.5.4);
- Superheated steam in a boiler (Sect. 2.5.5);
- Trailing centreline temperature in arc welding (Sect. 2.5.6).

These examples have been chosen because they concern processes that involve transport phenomena. Specially in the first two cases, there is a similitude with the dynamics of DCSFs. In each case it is explained how to choose the parameters that configure MUSMAR and the type of results that are obtained. The first example uses a laboratory plant, while examples 2 and 3 concern industrial plants.

When applying MUSMAR to a specific plant, the first step is to select the parameters that configure the controller in order to yield the best performance. This step is usually an iterative process in which one starts by an initial guess of what might be adequate parameters, makes simulations or plant experiments to improve each parameter and repeats this procedure until a satisfying configuration is obtained.

In order to configure MUSMAR the value of the following parameters must be chosen:

- *Minimum prediction horizon* T_1, in the cost (3.4). Set this to the value of the plant delay if this is known. Otherwise, set it to 1. In all the examples hereafter we choose $T_1 = 1$;
- *Prediction horizon* T, in the cost (3.4). Usually, there is a U-shaped dependency of the experimental cost on T. This can be understood in qualitative terms as follows: For small values of T the steady-state of the cost is not yet attained. For instance, if the process is nonminimum-phase, low values of T may not even stabilize the closed-loop. When T increases the steady-state controller gains are better approximated and this usually results in a performance improvement. However, when the horizon increases predictors become be less accurate and this causes a degradation of performance.
- *Forgetting factor* λ. Decreasing the forgetting factor makes the controller for agile to track changes in plant dynamics but increases the variance of the parameter

estimation error which in turn limits the best achievable performance. Due to the fact that MUSMAR uses a redundant description of the plant it is advantageous to use directional forgetting.

- *Number of samples in the pseudo-state* (3.33) of the output, n, and of the manipulated variable, m. In general, performance improves when the controller complexity $n + m$ increases. However, for constant $n + m$ a better performance is yielded when n increases. It is remarked that other variables can be included with advantage in the pseudo-state, for instance accessible disturbances or physical variables related to plant state. If this is the case, the number of delayed samples of these variables that is included in the pseudo-state also affects performance. For instance, if the variable is an almost constant disturbance that only presents low frequency changes in time, the best is to include just one sample. If, instead, the variable changes in a time scale similar to plant dynamics, more samples may be included and this has to be studied.
- *Penalty on the effort of the manipulated variable* ρ, in the cost (3.4). In general, decreasing ρ reduces the tracking error variance but increases the variance of the manipulated variable of of its variations, depending whether a positional or incremental algorithm is used. As shown in example 2, it is possible to increase ρ such that, with only a small degradation of tracking performance, there is a significant decrease in the variance of the manipulated variable, thus reducing the wearing of the actuator. It is remarked that, for large values of T so as to approximate the infinite LQ gains, the influence of rho on the poles of the closed-loop system can be studied by using the root-square locus (Lewis and Syrmos 1995). Another aspect to consider is the fact that, in the absence of integral effect, the tracking error will exhibit a steady-state offset that is bigger the bigger is the value of ρ.
- *Sampling period* T_s. The sampling period must be selected in accordance to the process dominant time constant. Too small sampling periods cause zeros of the discrete time model to migrate to the outside of the unit circle, thereby posing difficulties to the controller (for instance requiring that the horizon is increased). Too long sampling intervals may lead to an inadequate consideration of fast process modes and to inter-sample variance. Furthermore, one should bear in mind that during the sampling time the process is operating in open-loop. Thus, if a disturbance appears in-between two sampling instants, the controller will start to take actions to reject it only at the next sampling instant. Furthermore, the equivalent prediction horizon for the continuous-time dynamics is the product of the discrete-time horizon T and the sampling interval.
- *Parameters of dynamic weight filters*. These filters have influence in three aspects: Shaping the process output response to reference changes, high frequency noise filtering and robustness with respect to unmodeled dynamics.

These issues are now illustrated through the following examples as well as through the application to the control of temperature in a DCSF.

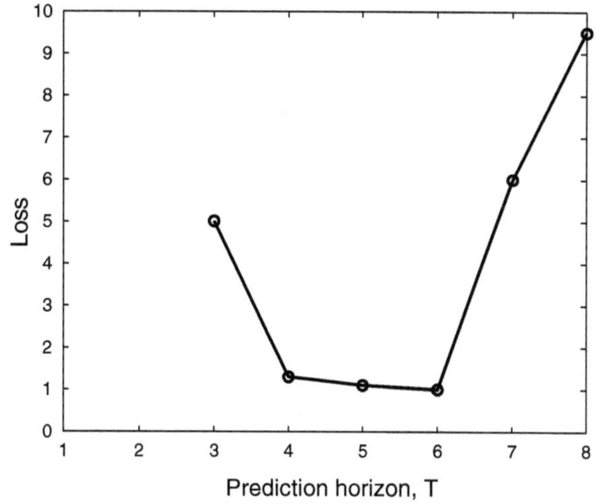

Fig. 3.6 Control of an air heating fan using MUSMAR. Influence of the choice of the prediction horizon on the steady-state loss

3.7.1 Example 1: Air Heating Fan

This example concerns the control of the temperature in a air hearting fan of the type described in Sect. 2.5.4, with the difference that the heating element is concentrated in a limited region of space, at the beginning of the pipe that drives the air. Hence, the plant dynamics includes a pure transport delay between the actuation and the plant output.

Figure 3.6 shows the dependence of the experimental loss obtained in steady-state for different values of the prediction horizon T. For low values of the prediction horizon ($T = 1$ and $T = 2$) the controller is not able to stabilize the closed-loop system, a fact due to the existence of a pure transport delay that demands a minimum value of T. For higher values of T there is a U-shaped dependency.

MUSMAR has been used with a prediction horizon $T = 5$, a penalty in the manipulated variable $\rho = 0.001$ and a dither noise variance $\sigma_\eta^2 = 0.01$. The structure of the pseudostate is defined by including $n_a = 3$ samples of the tracking error, $n_b = 1$ sample of the manipulated variable and $n_g = 1$ sample of the reference to track. The predictor parameters are estimated using RLS with directional forgetting, with an initial value of the covariance matrix of $P(0) = 1000$, and a directional forgetting factor of $\lambda = 0.005$. The sampling interval is 0.05 s.

Figure 3.7 shows a plot of the results obtained. The reference and temperature are plotted on the graphic of the top, the manipulated manipulated variable is plotted in the middle and 4 of the 5 controller gains are plotted below. All the plant input/output variable units are (V) (the units of the transducer), the range being between 0 and 5 V.

The evolution of the experiment protocol is as follows: after turning on the system, including the activation of MUSMAR, there is a sequence of steps of the reference that drive the air output temperature from a value corresponding to ambient temperature (0 V) to a value corresponding to 3.5 V. During this period, that lasts for about 1200

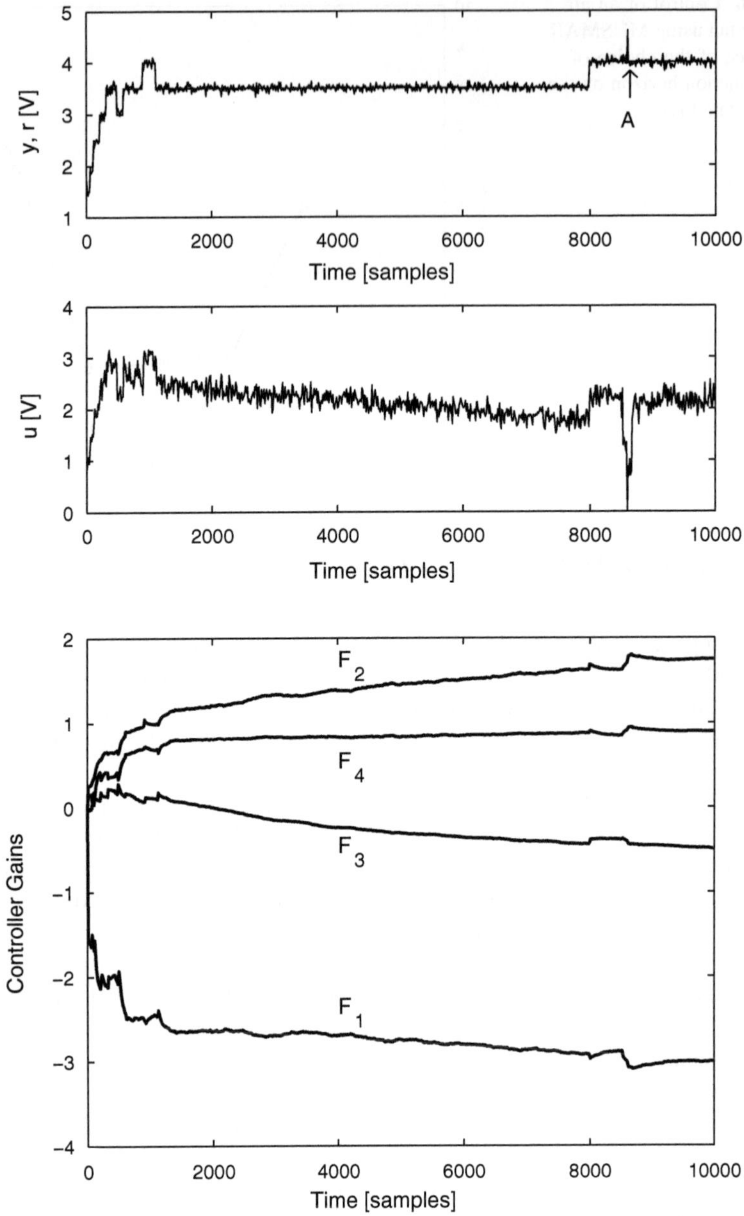

Fig. 3.7 Control of an air heating fan using MUSMAR. Reference and temperature (*above*), manipulated variable (*middle*) and controller gains (*below*)

samples, the gains converge after an initial adaptation transient. During the first five samples the gains are forced to be equal to a value that corresponds to an acceptable performance, but far from the optimum one.

The changes of the reference value help in the convergence of the controller gains. After this initial period, the reference is kept constant for a long period, up to sample 8000. During this period, although the tracking error power is kept approximately constant by the controller, there is a reduction of the average value of the manipulated variable, due to a slow adjustment of the controller gains. This happens because the pipe is slowly accumulating heat and, in this process, less and less heat is required to be released by the manipulator in order to keep the outlet air temperature at a constant value. In other words, this example illustrates the adaptation of the controller gains to a change in linearized process dynamics.

At sample 8000 there is a change in the reference value and, at sample 8500 (the time instant marked by the arrow labeled A), there is a sudden change in the position of the air admission register. This change is caused by an external factor and is included in order to exhibit the algorithm response to a rupture in dynamics, such as the one caused by a fault in a plant component. The manipulated variable is adjusted by feedback but, in addition, there is a readjustment in controller gains to meet the new plant dynamics. It is remarked that, in addition to a parameter change there is also a change of plant input/output transport delay because the flow of air is affected. MUSMAR is able to tackle this change in delay because of its structure based on a multistep cost function. The convergence speed of the gain re-settlement can be improved by manipulating the RLS covariance matrix (Costa and Lemos 1991).

3.7.2 Example 2: Superheated Steam in a Boiler

Example 2 concerns the temperature control of super-heated steam in the boiler unit described in Sect. 2.5.5. As explained in this section, the objective is to regulate the steam temperature at the outlet of a superheater by manipulating the command of the spray water injection valve position.

The prediction horizon was set at $T = 15$ discrete time instants and the sampling interval was $T_s = 5\,\text{s}$. This means that, in continuous time, the prediction horizon is $T \times T_s = 75\,\text{s}$. This value should be selected bearing in mind that the time scale of the plant, including the pure delay, varies with the value of steam flow. Indeed, since the water spray is injected at the input of the superheater (a set of parallel pipes through steam is made to pass in order to recover heat from the flue gas of the furnace) its influence at the output depends on the travel time which, in turn, is inversely proportional to steam flow. Hence, the higher the value of the flow, the faster becomes the response, with a smaller pure delay. For instance, if $T = 10$ and $T_s = 2\,\text{s}$ the controller yields acceptable results at the maximum steam flow of 150 t/h, but is unable to even stabilize the plant at the minimum flow of 100 t/h.

Furthermore, it should be kept in mind that there are some trade-offs in selecting. As already mentioned in the general remarks made in the begining of Sect. 3.7, the value sampling value must neither be decreased beyond a certain value, because

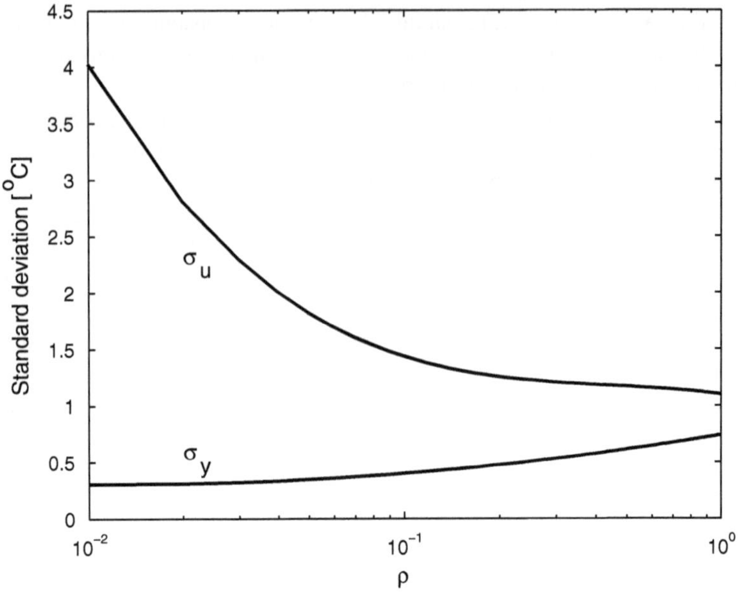

Fig. 3.8 Control of superheated steam temperature in a boiler unit with MUSMAR. Effect of the penalty parameter ρ on the standard deviations of the tracking error [°C] and of the valve position [%]

the resulting sampled dynamics may become nonminimum phase, nor be increased too much, because the plant is working in open loop between sampling intervals. Therefore, to encompass the possible outcomes of plant dynamic time scale due to changes of steam flow one must increase T. However, this increase has also a limit because, due to the loss of precision of predictors over large horizons, the overall performance also starts to degrade for high values of T.

In conclusion, one has to evaluate the possible ranges of T and T_s, by using a priori information about the plant and preliminary plant experiments, and optimize their values within them.

Another controller tuning knob is the value of the manipulated variable penalty, ρ that also forms an important way to configure the type of response of the controlled system. Figure 3.8 shows the effect of ρ on the standard deviations of the valve move σ_u (expressed in [%]) and of the tracking error σ_y (expressed in [^{o}C]). When ρ increases, σ_u decreases and σ_y increases. This means that there is a trade-of in the selection of ρ. Above a certain threshold of ρ the controller is in this case no longer able to stabilize the closed-loop.

In the example, the pseudostate contains $n_a = 3$ samples of the tracking error, $n_b = 2$ samples of the manipulated variable, 1 sample of the steam temperature at the input of the superheater and immediately after the water injection, 2 samples of the steam flow, 2 samples of the flow of air entering the furnace and 1 sample of a constant whose value is the average of the manipulated variable.

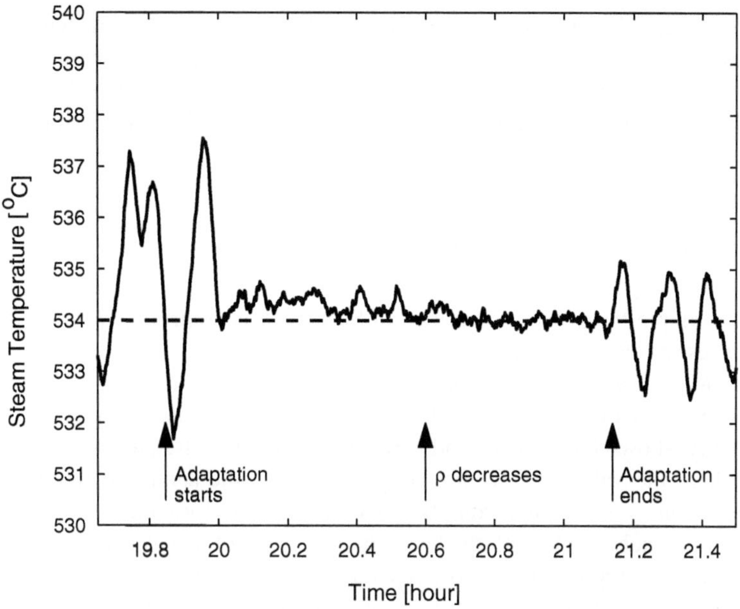

Fig. 3.9 Control of superheated steam temperature in a boiler unit with MUSMAR

The estimation of the predictor parameters is done with directional forgetting using a factorized algorithm and with a forgetting factor of $\lambda = 0.998$. The initial value of the covariance matrix is $P(0) = 1000I$. The standard deviation of the dither noise η was set to a value corresponding to 1 % of the maximum valve opening.

Figure 3.9 shows the record of an experiment performed using the above configuration of MUSMAR. Initially, the plant is operated by a controller made by a cascade of two PID controllers. The inner loop controls the steam temperature at the injection point and the outer loop controls the superheated steam temperature by manipulating the reference of the inner loop. In adition, there are feedforward effects from the steam flow and the flow of the air entering the furnace.

Although this controller has been optimized by the plant technical staff, when the adaptive controller is connected (at the instant marked by the arrow with the label "Adaptation stars"), the standard deviation of the tracking error is reduced by a factor of more than 3. This happens no matter the persistent disturbance caused by constant stochastic variations of stem flow between 145 t/h and 150 t/h, due to consumption by two industrial chemical plants, in addition to power production with a turbine. Initially, a value of $\rho = 0.05$ is used. When this parameter is further reduced to $\rho = 0.01$ (at the instant marked by the arrow labeled "ρ reduced"), the off-set between the average steam temperature and the reference is reduced since integral effect is turned off as well as the fluctuations around the mean value. This improvement in reference tracking has the reverse payback of implying an increase in the standard deviation of the valve command from 1,6 to 4 %, in accordance to Fig. 3.8.

When adaptation is turned off (at the instant marked by the arrow with the label "Adaptation stars"), the standard plant controller enters again in operation and the tracking performance degrades. Further details on the application of MUSMAR to this plant, including a relation of control performance with plant economic performance, can be seen in Silva et al. (2000).

3.7.3 Example 3: Arc Welding

In this example the process described in Sect. 2.5.6 is controlled with MUSMAR. Further results can be found in Santos et al. (2000).

MUSMAR predictors are identified using directional forgetting RLS with $\lambda = 0.98$. The weight on the penalty of the manipulated variable is chosen to be $\rho = 500$. Figure 3.10 shows the cost dependency on the sampling frequency. In the results shown here the value of 3 Hz has been chosen for the sampling frequency. The choice $n = m = 3$ leads to a control structure whose performance is of the order of the minimum achievable with the least possible value of $n + m$. For this value of n and m, Fig. 3.11 represents the cost dependence on the horizon T.

Finally, Fig. 3.12 shows a record of a temperature control experiment when using the above configuration applied to a workpiece consisting of a steel bar of 12 mm thickness in series with another bar of 6 mm thickness. This change of dimensions imply that the controller has to adapt its gains when passing from one part of the piece to the other, a fact that is apparent from the controller gains record in Fig. 3.12. In order to eliminate a steady-state tracking offset an integrator fed with the tracking error has been included in parallel with MUSMAR (Santos et al. 2000).

3.8 Solar Field Predictive Adaptive Control

Hereafter with describe some experimental results obtained with the predictive adaptive controller MUSMAR when applied to the DCSF described in Sect. 1.2.1 of Chap. 1. We start by considering the basic MUSMAR algorithm as presented in Sect. 3.6.1. Then, in order to eliminate undesirable high frequency oscillations that may be present in the manipulated variable, the algorithm modified by the inclusion of dynamic weights is used. Finally, taking advantage of the robustness of MUSMAR with respect to variations in plant input/output transport delay, cascade control such as depicted in the block diagram of Fig. 1.15 of Chap. 1 is exemplified.

3.8.1 The Basic Algorithm

We start by considering that the basic MUSMAR algorithm is interconnected to the DCSF such as shown in Fig. 1.13 of Chap. 1. The controller structure is defined by including in the pseudo-state (3.33) 3 samples of the tracking error and 2 samples of the manipulated variable. Furthermore, in order to incorporate feedforward effects,

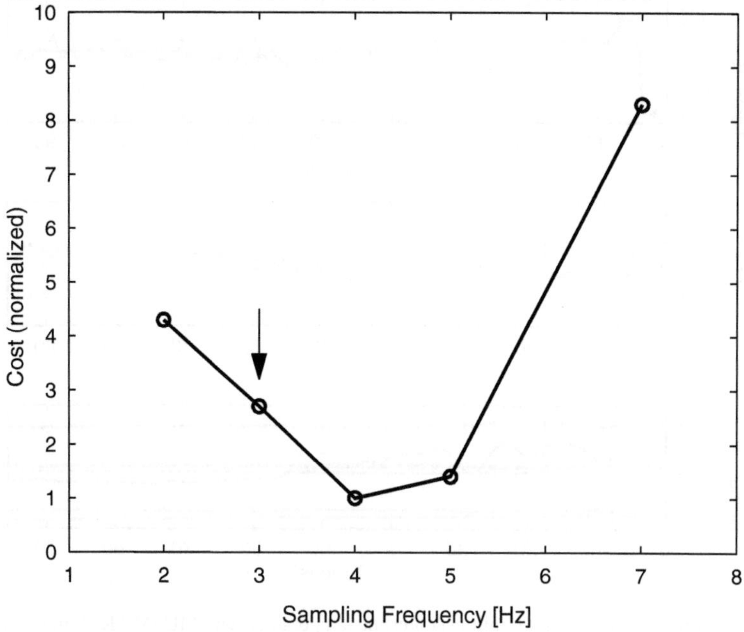

Fig. 3.10 Control of arc welding trailing centerline temperature with MUSMAR. Dependence of the cost on the sampling frequency

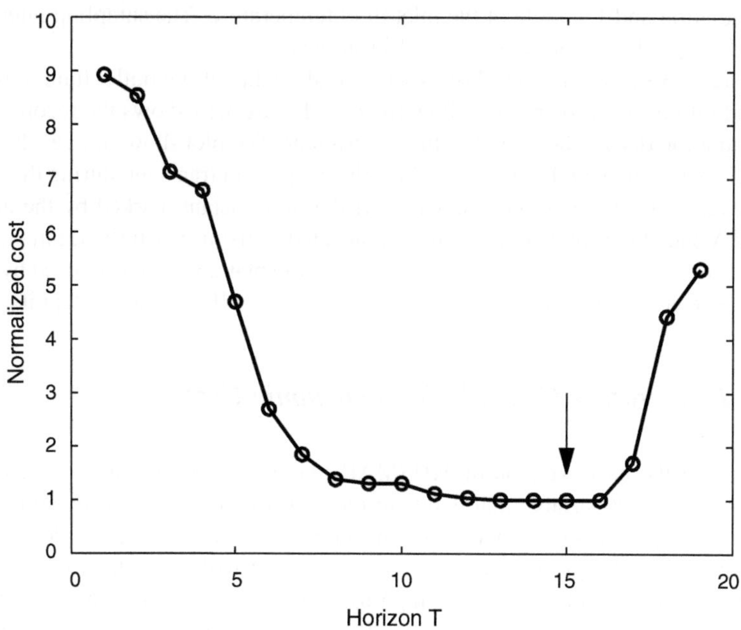

Fig. 3.11 Control of arc welding trailing centerline temperature with MUSMAR. Cost dependence on the horizon T for $n = m = 3$

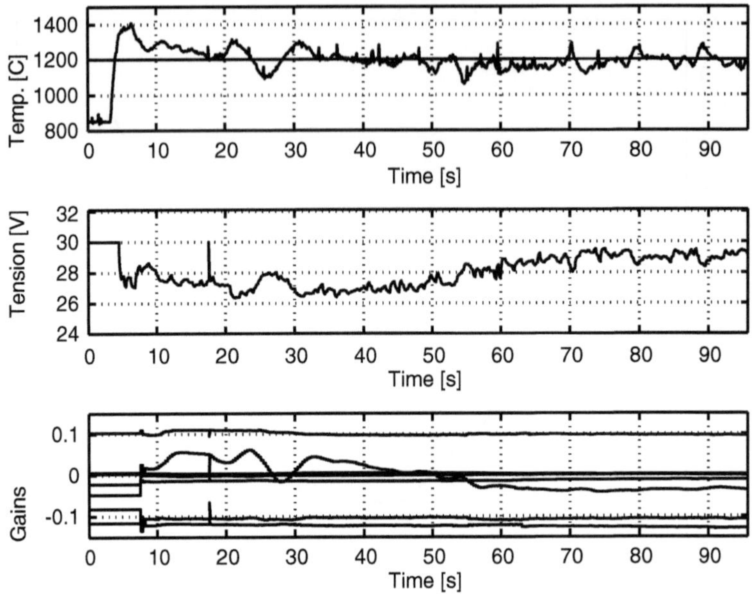

Fig. 3.12 Control of arc welding trailing centerline temperature with MUSMAR. Temperature and reference (*above*), electric tension (*middle*) and controller gains (*below*)

the pseudo-state includes also 1 sample of the reference to track, 1 sample of the solar radiation and 1 sample of the inlet fluid temperature. The sampling interval is 15s and the prediction horizon is $T = 15$ samples.

Figure 3.13 shows the record of experimental results of the outlet fluid temperature and reference (top) and fluid flow (below). Figure 3.14 shows the record of the solar radiation during the period of the experiment. The plot shows the results since MUSMAR is activated. In order to reduce the adaptation transient, during the initial 20 steps a set of constant gains is used. At the time instant marked by the arrows labeled A one of the collector loops is disconnected, causing a disturbance in the temperature as well as a change in the stating gain of the plant, which is reduced by 10 % (since there are 10 collector loops in parallel), causing MUSMAR to adapt its gains.

3.8.2 Temperature Control with a Dynamic Cost

When using the basic version of MUSMAR, there is a high frequency oscillation in the manipulated variable that is due to the high frequency DCSF dynamics that includes multiple anti-resonance peaks. In some cases, such as the example shown in Fig. 3.15 this high frequency oscillation can take the form of bursts that cause actuator wearing. As shown in Fig. 3.16, this high frequency oscillation is eliminated by the inclusion of dynamic weights in MUSMAR, according to the algorithm described in Sect. 3.6.4.

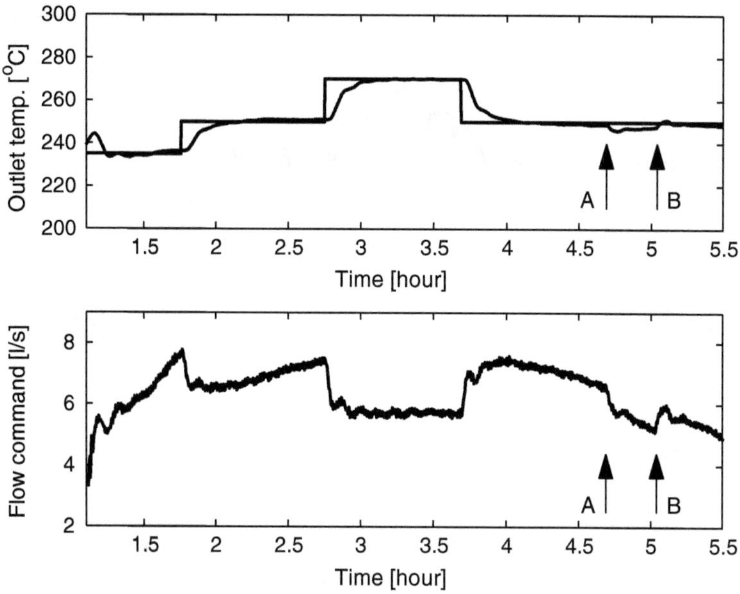

Fig. 3.13 Temperature control in a DCF with the basic MUSMAR algorithm. Outlet fluid temperature and reference (*top*) and fluid flow (*below*). Experimental results

3.8.3 Cascade Control

As explained in Chap. 1 the outlet of the collector loops is passed to a passive pipe and brought back to the storage tank, where it enters at the top. The passive pipe introduces a long delay that varies with the value of fluid flow. In a cascade control framework, the average of the temperatures at the outlet of the collector loops is taken as the intermediate variable. In this way, the dynamics is decomposed in two parts: The dynamics of the collector loops, that relates the fluid flow to the average of the temperatures at the collectors outlets, $\overline{T}$, and the dynamics of the pipe connecting the outlet of the loops with the inlet of the storage tank, that relates the temperatures in both points.

The former corresponds to the faster time constant, with a value of about 3 min. The latter corresponds to the slower part of the dynamics, made of a pure delay of about 9 min that varies with the fluid flow, in series with a time constant of about 2 min.

When using cascade control as in the block diagram of Fig. 1.15, the problems of rejecting disturbances in both subsystems are split apart.

Figure 3.17 shows experimental results obtained with cascade multi-rate MUSMAR control. In this experiment both T_1 and T_2 (the prediction horizons on both MUSMAR controllers in the cascade structure of Fig. 1.15) are made equal to 15 samples. The sampling intervals of both controllers are different, with the sampling interval of the inner controller $h_1 = 15$ s and the one of the outer controller $h_2 = 60$ s. When h_2 increases, the pure delay in P_2 measured in number of samples decreases.

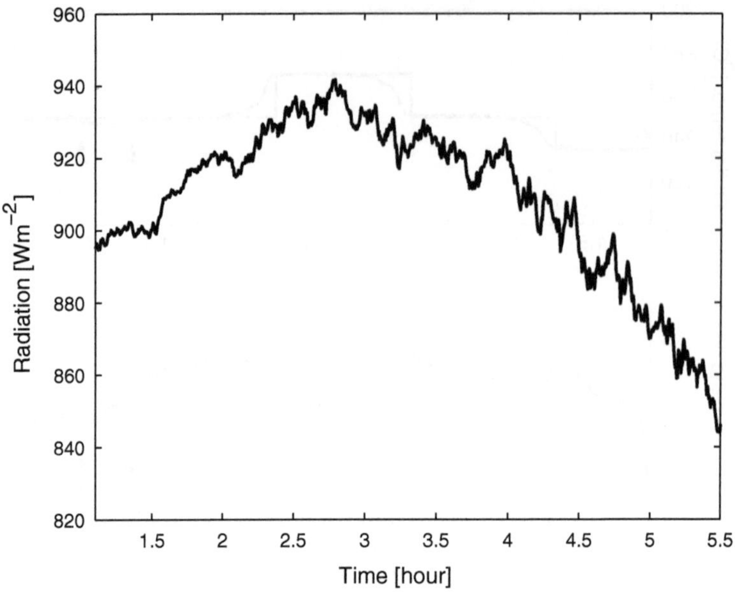

Fig. 3.14 The solar radiation in the experiment of Fig. 3.13

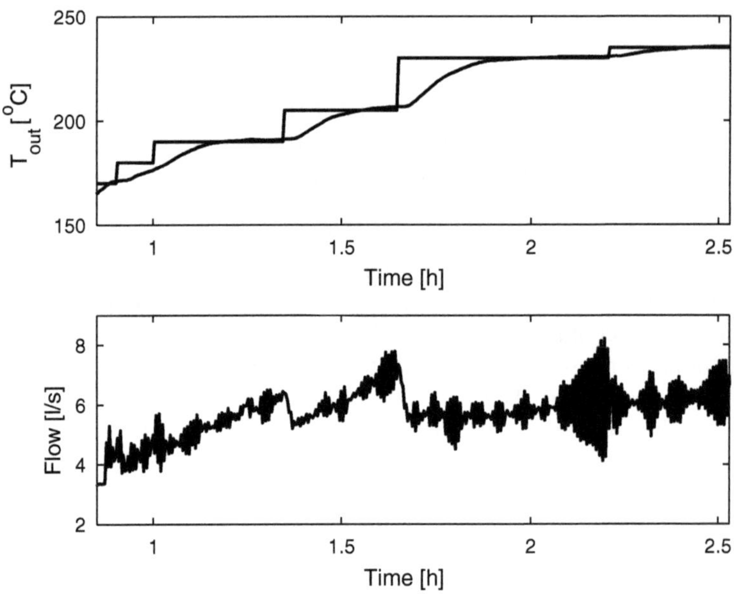

Fig. 3.15 Temperature control in a DCF with the basic MUSMAR algorithm in a situation where there are bursts of high frequency oscillation in the manipulated variable. Outlet fluid temperature and reference (*top*) and fluid flow (*below*). Experimental results

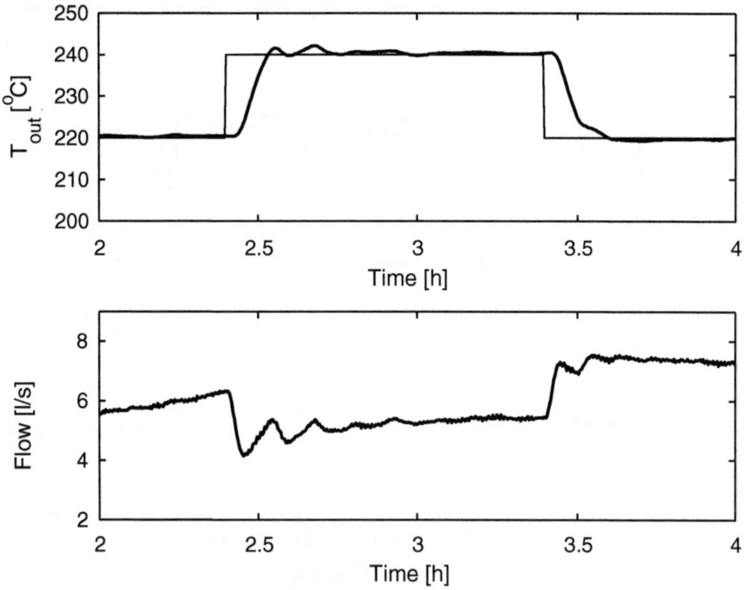

Fig. 3.16 Temperature control in a DCF with the MUSMAR algorithm modified to include dynamic weights. Outlet fluid temperature and reference (*top*) and fluid flow (*below*). Experimental results. Compare with the results in Fig. 3.15. The high frequency oscillation in the manipulated variable is eliminated by the inclusion of dynamic weights

The orders are set equal in both controllers and given by $n_i = 3$, $m_i = 2$. The control weight ρ_i is set equal to 0.001 in both controllers and the variance of the dither noise is 0.01.

The pure delay is apparent in Fig. 3.17, where the heat loss in the pipe connecting the outlet of the collector loops to the storage tank is also seen. The outer loop controller adjusts the set-point of the inner loop, such that y_1 is high enough to compensate for the thermal losses in the pipe and let the temperature at the inlet of the tank be equal to the set-point.

3.9 Dual Control and Start-Up

A major practical issue when controlling a plant whose dynamics is *a priori* unknown with an adaptive algorithm is the reduction of the start-up adaptation transient time. Many algorithms relies on the Certainty Equivalence Principle (CEP) (Åstrom 1987), meaning that the plant parameters are estimated on-line from plant data and the estimates are used to redesign the control law.

In this case, if the initial controller gains do not stabilize the plant, a strong burst at both the input and output signals may result, known as the adaptation transient.

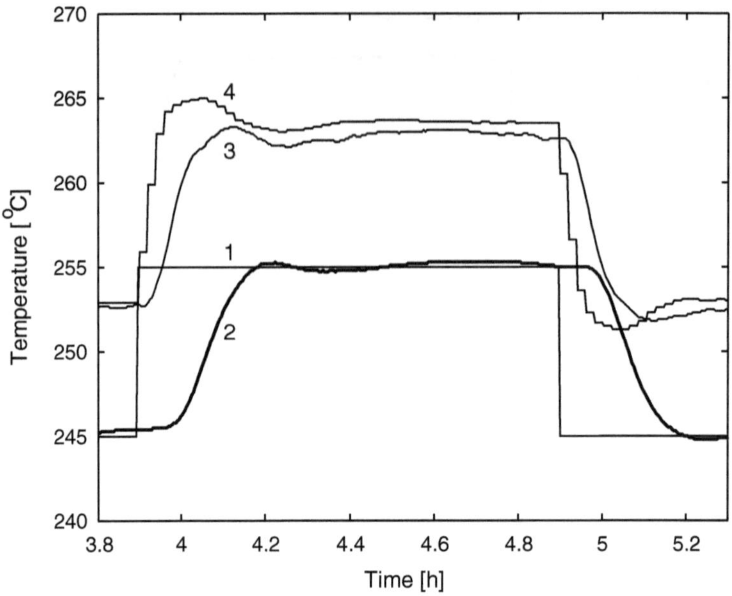

Fig. 3.17 Cascade control in a DCF with the MUSMAR algorithm. Experimental results

This instability effect can however help the controller in providing the adaptation mechanism with enough information to correctly identify the plant model. This compromise between the competing objectives of the control action (that tries to keep the plant state still) and plant model identification (that requires the plant state to be disturbed) is known as the "dual effect".

It can however happen that the start-up adaptation transient may be so strong as to prevent the practical use of the algorithm due to constraint violation. It may also drive the adaptation mechanism to a state where return to the optimum is very slow or even impossible due to the existence of local minima.

Therefore it is necessary to balance the dual actions of probing (system excitation in order to improve model estimation) and caution (reduction of excitation in order to improve control performance). Finding the adequate balance of the system excitation level so as to achieve a good control performance with adequate parameter excitation is the dual control problem.

The exact solution of the optimal dual control problem cannot be obtained exactly in general, neither analytically nor numerically. This is because of the growing dimension of the underlying space when time grows, in which alternate minimization and mean value evaluations have to be performed. Therefore, one must resort to approximate solutions of the solution for the dual problem.

Hereafter, the dual MUSMAR algorithm is obtained by modifying the basic MUSMAR predictive adaptive controller according to the bi-criteria approach (Filatov et al. 1997; Filatov and Unbehauen 2004).

3.9.1 The Bi-criteria Approach

The bi-criteria approach for the approximation of dual controllers relies on the consideration of two cost functionals at each sampling interval k

- the control cost, J_k^c and
- the uncertainty cost, J_k^a.

These two cost functionals correspond to the two objectives of dual control: controlling the process and generate an excitation such as to accelerate the convergence of parameter estimation. A dual adaptive controller based on the bi-criteria approach is obtained by the sequential minimization of both criteria according to the following steps (Filatov et al. 1997; Filatov and Unbehauen 2004) (see also Fig. 3.18 for an interpretation):

1. Obtain the cautious control component $u_c(k)$ by minimizing the control cost J_k^c:

$$u_c(k) = \text{argmin}_{u(k)} J_k^c. \tag{3.128}$$

2. Minimize the functional associated with uncertainty, J_k^a, in the interval Ω_k given by

$$\Omega_k = [u_c(k) - \vartheta(k); u_c(k) + \vartheta(k)], \quad \vartheta(k) \geq 0. \tag{3.129}$$

   This interval is centered in the cautious control component $u_c(k)$ and its length is proportional to the trace of the covariance matrix $P(k)$ of the least squares plant model identifier, meaning that

$$\vartheta(k) = \eta \, \text{tr}\{P(k)\}, \quad \eta \geq 0, \tag{3.130}$$

   where η is a parameter that adjusts the excitation level.

According to the bi-criteria procedure, the dual control law approximation, $u(k)$, actually applied to the plant at sampling interval k is given by

$$u(k) = \text{argmin}_{u(k) \in \Omega_k} J_k^a. \tag{3.131}$$

3.9.2 The Dual MUSMAR Algorithm

Consider a SISO ARMAX plant

$$A(q)y(k) = B(q)u(k) + C(q)e(k), \tag{3.132}$$

where $u(k)$ is the input; $y(k)$ is the output, $\{e(k)\}$ is a stationary zero-mean white sequence with variance σ_e^2 and A, B, C are polynomials in the unit forward shift

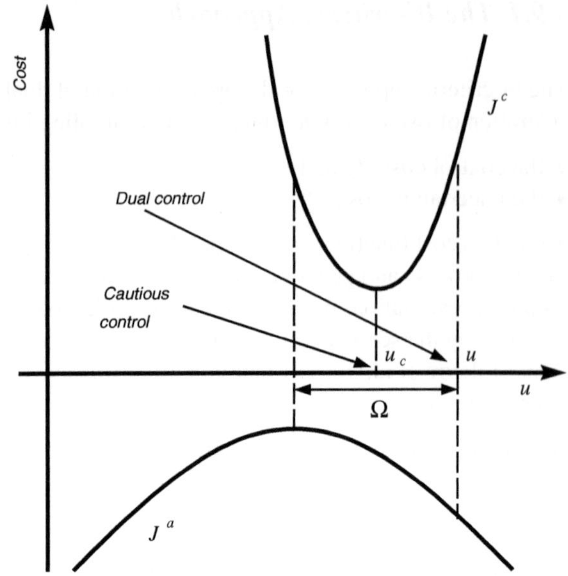

Fig. 3.18 Minimization for bi-criteria approximation of dual adaptive control

operator, q. Associated with (3.132) consider a quadratic cost functional

$$J_k^c = E\left\{\frac{1}{T}\sum_{i=1}^{T} y^2(k+i) + \rho u^2(k+i-1)|O^k\right\}, \tag{3.133}$$

where $\rho \geq 0$ is a design parameter that defines the penalty on the manipulated variable, $E\left\{\cdot|O^k\right\}$ stands for the expectation value conditioned on the $\sigma - algebra$ O^k generated by the information available up to time k and T is an integer design parameter corresponding to the prediction horizon. Consider also, a set of $2T$ least squares predictive models

$$\begin{aligned} \hat{y}(k+i) &= \hat{\theta}_i u(k) + \hat{\psi}_i^T s(k) \\ \hat{u}(k+i-1) &= \hat{\mu}_{i-1} u(k) + \hat{\phi}_{i-1}^T s(k) \\ i &= 1, \ldots, T \end{aligned} \tag{3.134}$$

The vector $s(k)$ is a sufficient statistic for computing in closed-loop the predictions $\hat{y}(k+i)$ and $\hat{u}(k+i)$ called the "pseudo-state" as in the basic version of MUSMAR, and the scalars and vectors $\hat{\theta}_i$, $\hat{\psi}_i$, $\hat{\mu}_{i-1}$, $\hat{\phi}_{i-1}$ are recursive least squares (RLS) estimates of the real unknown parameters at time k.

Let the parameters in equation (3.134) be considered as random variables θ_i, ψ_i, μ_i, ϕ_i $(i = 1, \ldots, T)$ with conditional mean $\hat{\theta}_i$, $\hat{\psi}_i$, $\hat{\mu}_i$, $\hat{\phi}_i$ and variances

$$\begin{aligned}
\sigma_{\theta i}^2 &= E\left\{(\theta_i - \hat{\theta}_i)^2 | O^k\right\} \\
\sigma_{\mu i}^2 &= E\left\{(\mu_i - \hat{\mu}_i)^2 | O^k\right\} \\
\sigma_{\theta\psi i} &= E\left\{(\theta_i - \hat{\theta}_i)(\psi_i - \hat{\psi}_i) | O^k\right\} \\
\sigma_{\mu\phi i} &= E\left\{(\mu_i - \hat{\mu}_i)(\phi_i - \hat{\phi}_i) | O^k\right\}.
\end{aligned} \tag{3.135}$$

As shown in Appendix E, solving this control problem results in the following cautious control law

$$u_c(k) = F^T s(k), \tag{3.136}$$

with the vector of controller gains F given by

$$F = -\frac{\sum_{i=1}^{T} \hat{\theta}_i \hat{\psi}_i + \rho \sum_{i=1}^{T} \hat{\mu}_{i-1} \hat{\phi}_{i-1} + \beta p_{us}}{\sum_{i=1}^{T} \hat{\theta}_i^2 + \rho \sum_{i=1}^{T} \hat{\mu}_{i-1}^2 + \beta p_{uu}}, \tag{3.137}$$

where the covariance matrix is

$$P = \begin{bmatrix} p_{uu} & p_{us}^T \\ p_{us} & p_{ss} \end{bmatrix} \tag{3.138}$$

and $\beta = \sum_{i=1}^{T} \sigma_{yi}^2 + \rho \sigma_{ui}^2$, σ_{yi}^2 and σ_{ui}^2 being the variance of the prediction errors of y_i and u_i as defined by Eq. (3.134). If β is made equal to zero, Eq. (3.137) reduces to the basic MUSMAR algorithm described in Sect. 3.6.1.

The first modification in the MUSMAR algorithm incorporates a cautious mechanism on the gain computation. During the startup of the algorithm the value of p_{uu} is large implying that the control action will be very small.

The second modification introduced in the basic algorithm concerns the plant excitation. The basic MUSMAR controller adds to the controller action a low power dither noise η with variance σ_η^2 to maintain the persistent excitation of the estimator. Instead of using constant power white dither noise added to the control input, the dual version uses a second criterion for acceleration of the parameter estimation process by maximizing the predictive errors value. This criterion consists in minimizing the second cost functional given by

$$J_k^a = -E\left\{\frac{1}{T}\sum_{i=1}^{T}\left[y(k+i) - \hat{y}(k+i)\right]^2 + \alpha\left[u(k+i-1) - \hat{u}(k+i-1)\right]^2 | O^k\right\}, \tag{3.139}$$

with $\alpha > 0$. As explained above the aim is to find the minimum of Eq. (3.139) in an interval Ω_k around the cautious control action defined by Eq. (3.137). As such, the control action is given by Eq. (3.131), with Ω_k according to Eq. (3.129), where the width of the interval is proportional to the trace of the covariance matrix defined in Eq. (3.130) and defines the amplitude of the excitation as shown in Appendix E. The control law obtained by the minimization of (3.139) is given by (see the Appendix E)

$$u(k) = u_c(k) + \vartheta(k)\,\mathrm{sign}\left\{p_{uu}u_c(k) + p_{us}^T s(k)\right\}. \tag{3.140}$$

To sum up, the algorithm is implemented as follows:

3.9.2.1 Dual MUSMAR Control Algorithm

At the beginning of each sampling interval performed, the following steps are given in a recursive way:

Step 1 Using RLS, update the covariance matrix, $P(k)$, and the estimates of the parameters in the following set of predictive models

$$\hat{y}(k - T + i) = \hat{\theta}_i u(k - T) + \hat{\psi}_i^T s(k - T), \tag{3.141}$$

$$\hat{u}(k - T + i - 1) = \hat{\mu}_{i-1} u(k - T) + \hat{\phi}_{i-1}^T s(k - T),$$

$$i = 1, \ldots, T.$$

$$s(k) = [y(k) \ldots y(k - n_a + 1)\, u(k - 1) \ldots u(k - n_b)\, r(k)]^T. \tag{3.142}$$

Step 2 Using the past prediction errors for each model, compute the value of β and use it to update the cautious control action as in equation (3.137).

Step 3 Apply the control given by equation (3.140) to the process manipulated input.

3.9.3 *Dual MUSMAR Control of a DCSF*

In the experimental results reported in Sect. 3.8, the startup procedure was performed with some "reasonable" initial gains obtained by simulations in a plant model or from past experiments. If no such a priori information is available, there can be violent adaptation transients that result in poor performance. Figure 3.19 (top) shows experimental results in such a situation, in which standard MUSMAR is used in a DCSF field with poor a priori information (resulting from a very crude simulation plant model).

On the other way, Fig. 3.19 (bottom) presents the output temperature and the reference to track when dual MUSMAR is used. The structure used for the pseudo-state is: 3 output terms ($n_a = 3$), 2 input terms ($n_b = 2$), the actual reference signal (1 term), the actual radiation measure (1 term) and the actual inlet temperature measure (1 term). This choice totalizes 8 terms, meaning that 8 (feedback and feed-forward) gains are tuned. The prediction horizon in this experiment is $T = 15$, with the sampling period being $T_s = 15$ s. The initial adaptation transient is avoided and the performance is clearly better with respect to the results of basic MUSMAR, shown in Fig. 3.19 (top).

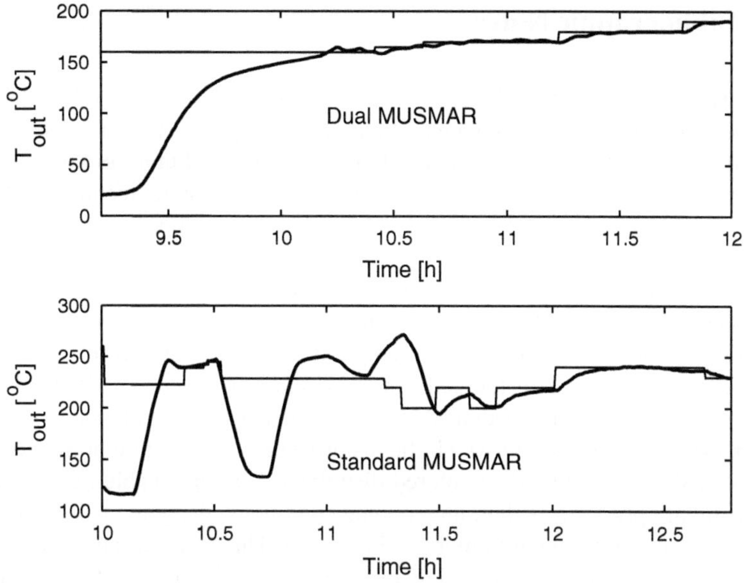

Fig. 3.19 Adaptation transient in the DCSF used with standard MUSMAR (*below*) and with Dual MUSMAR (*above*) with no a priori information about the process dynamics is used

3.10 Main Points of the Chapter

Temperature control of DCSFs raises problems that can be tackled by MPC. The difficulties to overcome include varying input–output transport delay and significant un-modeled dynamics that includes high frequency peaks, and is inherent to a finite dimension approximation of the exact infinite dimension model of DCSFs. When using MPC based on linear models, adaptation allows to adjust the controller gains to the working point as well as to other sources of uncertainty. Several MPC algorithms based on linear models have been successfully used to control DCSFs, two main examples being GPC and MUSMAR. Both of these controllers minimize a multistep quadratic cost defined over a time horizon that slides with the current time.

The basic adaptive MPC that rely on linear models can be subject with advantage to several modifications. The inclusion of dynamic weights allow to shape the controlled system frequency response and therefore, if properly used, to enhance the stability robustness properties of the algorithm. Another significant modification consists of on-line modification of the weights that penalize the manipulated variable to asymptotically satisfy constraints on its average value. Finally, start-up adaptation transients can be reduced either by a proper initialization of the adaptation algorithm (including not only the value of the initial plant model parameter estimates but also its error covariance matrix) or by using dual version of the algorithms.

3.11 Bibliographic Notes

The first attempt to design a self-tuner was described in Kalman (1958). The algorithm proposed combined the least squares estimation of the parameters of a deterministic linear difference equation with a control law to be selected in accordance with the application considered in each case. Lack of adequate computer technology limited the impact of these ideas, although a dedicated analog computer was build at the time to test the controller.

The successive step consisted in the the inclusion of stochastic disturbances in the model and the formulation of the control problem as the minimization of the plant output variance at the earliest sample affected by the current manipulated variable move, resulting in the self-tuning regulator (Åstrom 1973). In the self-tuning regulator adaptation is embedded by estimating model parameters with RLS (Ljung 1983). In the cases in which the disturbance is not a white noise stochastic process (a situation referred as colored disturbance), a remarkable fact, known as the self-tuning property, is that, under some conditions, the gains will converge to the ones that minimize the minimum variance cost. This fact is established (Åstrom et al. 1977) using the so called ODE method (Ljung 1977) according to which a deterministic ordinary differential equation is associated to the difference stochastic equation to be studied. In rough terms, the solution of the ODE corresponds to the average behavior of the solutions of the difference equation, the equilibrium points being the same. Applying the ODE method to the "standard" self-tuner (RLS model estimation plus minimum variance cost) leads to the conclusion that, provided the closed loop is stable and under a positive real condition, the controller gains converge globally to the gains that minimize (Ljung 1977).

Since the minimum variance control law yields a closed-loop that cancels the zeros of the open-loop system, for nonminimum-phase plants there will be unstable modes. Even when the closed-loop is stable, minimum variance control is too reactive, yielding an excessive control action. In order to overcome the limitations of the minimum variance control law, a new version of the basic self-tuning was proposed in which the cost is modified by the inclusion of a term that penalizes the manipulated variable (Clarke and Gawthrop 1975, 1979), the so called detuned minimum variance control. Detuned minimum variance control was an important step forward because the extra penalty regularizes the control variable, providing the designer with a knob to change the type of closed-loop response obtained.

The resulting control law can be used to stabilize open-loop stable nonminimum phase systems but, as already recognized in the original work (Clarke and Gawthrop 1975, 1979), is not able to stabilize a plant that is simultaneously open-loop unstable and nonminimum phase. For these type of plants, one should resort to a control law based on a cost function defined over an extended horizon. Hence, the idea arose of designing the control by minimizing an extended horizon cost that approximates the steady-state quadratic cost. Together with the concept of receding horizon minimization, this approach leads to several MPC algorithms based on linear models (Richalet et al. 1978; Keiser and Vauwenberghe 1981; Rouhani and Mehra 1982; Peterka 1984; Greco et al. 1984; Sánchez and Shah 1984; Clarke et al. 1987a, b).

An issue that raised a strong controversy in the end of the 1980s was the problem of finding a value of the horizon T that ensures stability. This lead to a number of modifications with various techniques (Mosca et al. 1990; Clarke and Scattolini 1991; Mosca and Lemos 1992; Bitmead et al. 1990).

Several modifications and extensions were also considered, including the multivariable case (Sánchez and Shah 1984; Shah 1989; Scattolini and Schiavoni 1992; Mohtadi 1992), state-space modeling (Li 1989; Elshafei 1995) and soft (i.e., depending on the mean value of the variables) constraints (Mosca 1992).

When hard constraints are included, the controller may no longer be expressed as a linear feedback of a (possibly nonminimal) state and one has to resort to nonlinear programming numerical algorithms to find the optimal value of the manipulated variables (Maciejowski 2002). Recent formulations (Kansha and Chiu 2009) allow to handle constraints in an approximate, although systematic, way. Other algorithms that incorporate hard constraints in adaptive MPC using various approaches are found in Adetola et al. (2009), Kim and Sugie (2008), Kim et al. (2008), Maniar et al. (1997), Nikolakopoulos et al. (2006), Tanaskovic et al. (2013), Veres and Norton (1993).

Filtering and the inclusion of dynamic weights in one and two-degrees of freedom controllers to enhance robustness is treated for GPC in Yoon and Clarke (1995) and for MUSMAR in Coito et al. (1997).

A comprehensive exposition of adaptive dual control is made in Filatov and Unbehauen (2004). As discussed in Filatov and Unbehauen (2000), the difficulties in obtaining exact dual algorithms led to a number of simplified approaches, of which (Chan and Zarrop 1985; Filatov et al. 1997; Lindoff et al. 1999; Milito et al. 1982) are examples. An adaptive predictive controller with dual features is reported in Lindoff et al. (1999). It results from the modification of an optimal multi-step stochastic quadratic cost by replacing the unconditioned mean with a mean conditioned on the σ-algebra generated by the observations available up to current time. Another approach solves a bi-criteria optimization problem (Filatov et al. 1997) and yields an acceptable computational load. A number of industrial type applications of dual adaptive control are reported in the literature. In Ismail et al. (2003), the application of an adaptive dual control strategy to a bent blade paper coating machine is reported. Similar process characteristics motivate also embodying dual features in controller as described in Allison (1995), where probing is mentioned to be a useful way of starting up the controller. The bi-criteria approach has been applied in Filatov (1998) to the speed control of a thyristor driven DC-motor. The derivation of the dual MUSMAR algorithm and its application to a DCSF is presented in Silva et al. (2005).

References

Adetola V, DeHan D, Guay M (2009) Adaptive model predictive control for constrained nonlinear systems. Systems and Control Letters 59:320–326

Allison B, Ciarniello J, Tessier P, Dumont G (1995) Dual adaptive control of chip refiner motor load. Automatica 31(8):1169–1184

Åstrom KJ, Borisson U, Ljung L, Wittenmark B (1977) Theory and applications of self-tuning regulators. Automatica 13:457–476

Åstrom KJ, Wittenmark B (1973) On self-tuning regulators. Automatica 9:185–199

Åstrom KJ, Wittenmark B (1997) Computer controlled Systems. Prentice Hall, New Jersey

Åstrom KJ (1987) Adaptive feedback control. Proc IEEE 75(2):185–217

Bierman GJ (1977) Factorization methods for discrete sequential estimation. Academic Press, New York

Bitmead RR, Gevers M, Wertz V (1990) Adaptive optimal control. Prentice Hall, New Jersey

Chan S, Zarrop M (1985) A suboptimal dual controller for stochastic systems with unknown parameters. Int J Control 41:507–524

Clarke DW, Mohtadi C, Tuffs PS (1987a) Generalized predictive control—part II. Extensions and interpretations. Automatica 23(2):149–160

Clarke DW, Mohtadi C, Tuffs PS (1987b) Generalized predictive control—part I. The basic algorithm. Automatica 23(2):137–148

Clarke DW, Scattolini R (1991) Constrained receding horizon predictive control. Proc IEE 138-D:347–354

Clarke DW, Gawthrop PJ (1975) Self-tuning controller. Proc IEE-D 122:929–934

Clarke DW, Gawthrop PJ (1979) Self-tuning control. Proc IEE-D 126:633–640

Coito F, Lemos JM, Silva RN, Mosca E (1997) Adaptive control of a solar energy plant: exploiting accessible disturbances. Int J Adapt Control Signal Process 11: 327–342

Costa BM, Lemos JM (1991) A long-range adaptive controller for time-varying plants with sudden changes. Int J Adapt Control Signal Process 5:219–230

Cutler CR, Ramaker BL (1980) Dynamic matrix control: a computer control algorithm. In: Proceedings of the joint American control conference, San Francisco, USA

De Keiser R, Van Vauwenberghe A (1981) A self-tuning multistep predictor application. Automatica 17:167–174

Elshafei A, Dumont G, Elnaggar A (1995) Stability and convergence analysis of an adaptive GPC based on state-space modelling. Int J Control 61(1):193–210

Filatov NM (1998) Application of adaptive dual controllers to a DC-motor. IEE Proc Control Theory Appl 145(3):299–305

Filatov NM, Unbehauen H (2000) Survey of adaptative dual control methods. IEE Proc Control Theory Appl 147(1):118–128

Filatov NM, Unbehauen H (2004) Adaptive dual control. Springer, Berlin

Filatov NM, Unbehauen H, Keuchel U (1997) Dual pole placement controller with direct adaptation. Automatica 33:113–117

Greco C, Menga G, Mosca E, Zappa G (1984) Performance improvements of self-tuning controllers by multistep horizons. The MUSMAR approach. Automatica 20:681–699

Ismail A, Dumont G, Backstrom J (2003) Dual adaptive control of paper coating. IEEE Trans Control Syst Tech 11(3):289–309

Kalman RE (1958) Design of a self-optimizing control system. Trans ASME 80:468–478

Kansha Y, Chiu M (2009) Adaptive generalized predictive control based on JITL technique. J Process Control 19:1067–1072

Kim T-K, Sugie T (2008) Adaptive recedinh horizon predictive controlfor constrained discrete-time systems with parameter uncertainties. Int J Control 81:62–73

Kim J-S, Yoon T-W, Shim H, Seo JH (2008) Switching adaptive output feedback model predictive control for a class of input constrained linear plants. IET Control Theory Appl 2:573–582

Kulhavý R (1987) Restricted exponential forgetting in real-time identification. Automatica 23(5):589–600

Kwon WH, Han S (2005)Receding horizon control. Springer, Berlin

Lewis FL, Syrmos VL (1995) Optimal control. Wiley, New York

Li S, Lim K, Fisher D (1989) A state-space formulation for model predictive control. AICHE J 35:241–249

Lindoff B, Holst J, Wittenmark B (1999) Analysis of approximations of dual control. Int J Adapt Control Signal Process 13(7):593–620
Ljung L (1977). On positive real transfer functions and the convergence of some recursive schemes. IEEE Trans Autom Control AC-22(4):539–551
Ljung L, Soderstrom T (1983) Theory and practice of recursive identification. The MIT Press, Cambridge
Maciejowski JM (2002) Predictive control with constraints. Prentice Hall, New Jersey
Maniar V M, Shah SL, Fisher DG, Mutha RK (1997) Multivariable constrained adaptive GPC: theory and experimental evaluation. Int J Adapt Control Signal process 11: 343–365
Milito R, Padilla C, Padilla R, Cadarin D (1982) An innovations approach to dual control. IEEE Trans Autom Control AC27:132–137
Mohtadi C, Shah SL, Fisher DG (1992) Frequency response characteristics of MIMO GPC. Int J Control 55(4):877–900
Mosca E (1995) Optimal, predictive, and adaptive control. Prentice Hall, New Jersey
Mosca E, Zappa G (1989) ARX modeling of controlled ARMAX plants and LQ adaptive controllers. IEEE Trans Autom Control 34(3):371–375
Mosca E, Zappa G, Lemos JM (1989) Robustness of multipredictor adaptive regulators: MUSMAR. Automatica 25(4):521–529
Mosca E, Lemos JM (1992) A semi-infinite horizon LQ self-tuning regulator for ARMAX plants based on RLS. Automatica 28:401–406
Mosca E, Lemos JM, Mendonça TF, Nistri P (1992) Adaptive predictive control with mean-square input constraint. Automatica 28(3):593–598
Mosca E, Lemos JM, Zhang J (1990) Stabilizing I/O receding-horizon control. In: Proceedings of 29th IEEE conference on decision and control, Honolulu, pp 2518–2523
Nikolakopoulos G, Dritsas L, Tzes A, Lygeros J (2006) Adaptive constrained control of uncertain ARMA-systems based on set membership identification. In: Mediterranean conference on control and automation, Ancona, Italy
Peterka V (1984) Predictor-based self-tuning control. Automatica 20(1):39–50
Richalet J, Rault A, Testud JL, Papon J (1978) Model predictive heuristic control: application to industrial processes. Automatica 14:413–428
Rouhani R, Mehra RK (1982) Model algorithmic control (MAC); basic theoretical properties. Automatica 18(4):401–414
Sánchez JM, Rodellar J (1996)Adaptive predictive control. Prentice-Hall, New Jersey
Sánchez JMM, Shah SL (1984) Multivariable adaptive predictive control of a binary destilation column. Automatica 20(5):607–620
Santos TO, Caetano RB, Lemos JM, Coito FJ (2000) Multipredictive adaptive control of arc welding trailing centerline temperature. IEEE Trans Control Syst Tech 8(1):159–169
Scattolini R, Schiavoni N (1992) A note on multivariable generalized predictive control. Int J Control 56(1):253–257
Shah SL, Mohtadi C, Clarke DW (1989) Multivariable adaptive control without a prior knowledge of the delay matrix. Syst Control Lett 9:295–306
Silva RN, Filatov N, Hunbehauen H, Lemos JM (2005) A dual approach to start-up of an adaptive predictive controller. IEEE Trans Control Syst Tech 13(6):877–883
Silva RN, Shirley P, Lemos JM, Gonçalves AC (2000) Adaptive regulation of super-heated steam temperature: a case study in an industrial boiler. Control Eng Pract 8:1405–1415
Tanaskovic M, Fagiano L, Smith R, Goulart P, Morari M (2013) Adaptive model predictive control for constrained linear systems. In: 12th European Control conference, Zurich, Switzerland, pp 382–387
Veres SM, Norton JP (1993) Predictive self-tuning control by parameter bounding and worst case design. Automatica 29(4):911–928
Yoon Y, Clarke DW (1995) Observer design in receding-horizon predictive control. Int J Control 61(1):171–192

Lindoff B, Holst J, Wittenmark B (1999) Analysis of approximations of dual control. Int J Adapt Control Signal Process 13(7):593–620
Ljung L (1977) On positive real transfer functions and the convergence of some recursive schemes. IEEE Trans Autom Control AC-22(4):539–551
Ljung L, Soderstrom T (1983) Theory and practice of recursive identification. The MIT Press, Cambridge
Maciejowski JM (2002) Predictive control with constraints. Prentice Hall, New Jersey
Mathur VM, Shah SL, Fisher DG, Mutha RK (1997) Multivariable constrained adaptive GPC: theory and experimental evaluation. Int J Adapt Control Signal process
[illegible]

Chapter 4
Multiple Model Adaptive Control

When considering methods that can be applied to a wide class of operating regimes, an alternative to the adaptive controllers described in Chap. 3 consists in the use of algorithms that rely on multiple models. In general terms, to use this type of methods the plant dynamics is assumed to be unknown, and possibly slowly time varying, but amenable to be represented by one of the models in a a priori defined set of models (the model bank). An important case is when the plant is linear and its dynamics depends on a parameter, being denoted as the *Linear Parameter value* (LPV) class of models Balas (2002), Mohammadpour and Scherer (2010).

In classical gain-scheduling, to each one of the possible outcomes of the plant model a controller is designed in such a way that, if the controller is coupled with the corresponding model (or with the models in a subset of the overall model set that are similar in an adequate sense), then the resulting closed-loop satisfies the specifications. The adaptation mechanism consists thus of an algorithm that, given the observations of the input, output, and/or possibly other plant signals, decides at each sampling interval what model that best fits the observed plant dynamical behavior. In the more powerful class of LPV-based design methods controller design is formulated as a convex optimization problem with linear matrix inequality (LMI).

The simplest way to select the controller that should be activated is by means of a scheduling variable, according to a procedure called *gain scheduling*. If the dynamics of the plant depends on a variable, or set of variables, that define its operating point, the controller is selected as the entry of a table whose indexes are a function of the value read for those variables. As an example, consider a DCSF whose temperature is controlled by a PID as in Fig. 1.16. If the PID gains are tuned to yield a good response in an intermediate range of temperature values, as shown in Fig. 1.16, the response will be sluggish at low temperatures and too oscillatory, or even unstable, at high temperatures. According to the classical gain scheduling approach, one can consider three sets of gains for the PID, that are optimized in three different temperature ranges corresponding to low, intermediate, and high values of temperature. Therefore, each set of gains will be used depending on the temperature measure at each sampling interval.

J. M. Lemos et al., *Adaptive Control of Solar Energy Collector Systems*,
Advances in Industrial Control, DOI: 10.1007/978-3-319-06853-4_4,

Although gain scheduling proved to be very useful in many industrial applications it has the obvious limitation of being unable to adapt the controller when the plant dynamics changes due to causes that were not foreseen in anticipation and do not depend on the value of the indicator variables. This way of operation lead gain scheduling to be sometimes referred as "open-loop adaptation". This limitation is overcome by supervised multiple model adaptive control (SMMAC).

In SMMAC all the models in the model bank are excited by the same manipulated variable applied to the plant input. Their outputs are compared with the actual plant output and the one that yields the least error is considered to be the best plant representation. The corresponding controller is then used to generate the manipulated variable value that is actually applied to the plant input.

When considering these adaptive control structures, several issues have to be considered:

- Since these methods rely on switching among different candidate controllers, one has to ensure bumpless transfer when switching between the different sources of the manipulated variable;
- Conditions for stability have also to be ensured. Indeed, it is not enough to guarantee stability of the overall system to ensure that each controller stabilizes the model that elects it since it is possible to switch between stable systems in a way that results in an unstable time varying system;
- The local controllers have to be designed such as to be robust with respect to unmodeled plant dynamics;
- Noise and disturbances have to be taken into account when designing both the local controllers and the supervisor that perform switching.

In the present chapter, we address control based on multiple model algorithms. We start by introducing classical gain-scheduling and then we progress to SMMAC. Experimental results with two different types of plants are presented to illustrate the methods. First, we consider the air-heating fan described in Sect. 2.5.4 for which detailed experimental results that illustrate several important design issues are presented. Then, the application to a DCSF is considered.

4.1 Controller Structure

In supervisory based multiple model control, a supervisor decides which controller, among a number of a priori designed controllers $C_1, \ldots, C_N$ is to be connected with the plant to control. Hereafter, we designate the C_i as *local controllers* and their ensemble as the *model bank*. Figure 4.1 represents a block diagram of this type of adaptive control. When the plant input increment Δu is connected in permanence to one of the outputs of the candidate controllers, a closed loop system is obtained. It is assumed that the plant can be adequately controlled by one of the controllers, C_{i^*}, the problem being to find the value of the index i^*.

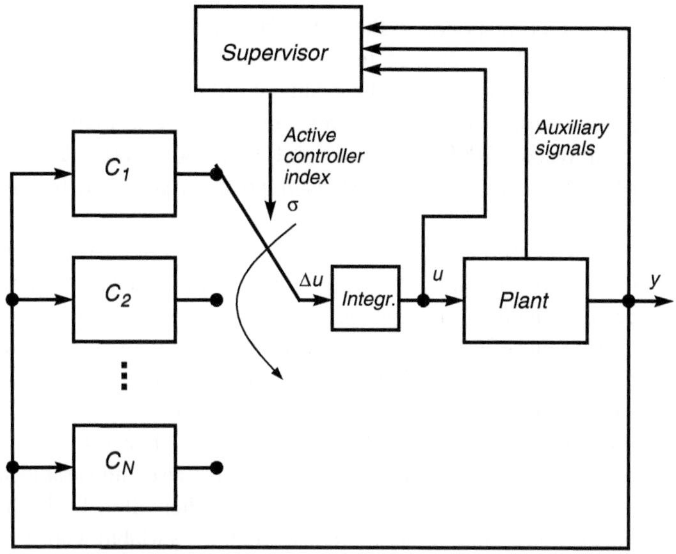

Fig. 4.1 Supervised control

In this chapter, we assume that the plant is linear and may change its dynamics only through jumps of a vector of parameters that define its poles, zeros, and static gain. Furthermore, it is assumed that these jumps occur only after intervals of time in which the plant is time-invariant, that are long compared to the plant dominant time constant. In addition, it is assumed that, for each of the possible outcomes of plant dynamics, there is at least one controller that stabilizes the closed-loop and yields the performance that is specified.

In practice, this assumption implies that each controller is able to meet the specifications for a certain range of possible plant dynamics, and that the controller bank covers all the possibilities. One may also consider the problem of enlarging the model bank with new controllers when the supervisor concludes that none of the existing controllers is adequate (Lourenço and Lemos 2008, 2006), a situation that is not discussed here.

In the diagram of Fig. 4.1, switching among controllers happens mainly in three situations. In the first place, when the system is started-up, and unless there is some a priori knowledge about the plant dynamics, the controller employed for the first time is likely to yield a poor performance. This adaptation transient is perceived by the supervisor, that will start to look for the best controller, using the algorithms that are explained below in this chapter. When the "best" controller is connected, if the plant varies, the supervisor will switch to a new controller if it is expected to yield a superior performance. Finally, even if a controller is the "best" one, there can be switchings induced by noise and disturbances.

Switching corresponds thus to adapting the controller to the plant dynamics with the ultimate goal of improving performance, but it also causes some undesirable effects. The first one corresponds to the fact that switching from one controller to another may cause an abrupt change in the manipulated variable u, that degrades performance and may be harmful to the plant. Avoiding this type of effects is the realm of *bumpless transfer*, for which several methods are available (Hanus et al. 1987; Turner and Walker 2000; Zaccarian and Teel 2002, 2005). In the methods considered in this book, the bumpless transfer problem is solved by inserting an integrator that is common to all controllers, before the plant manipulated input, as shown in Fig. 4.1. Therefore, the local controllers generate only increments Δu of the manipulated variable. Even if Δu changes abruptly, the integrator output, that is directly connected to the plant manipulated input, is always continuous.

The other undesirable effect of switching is the possibility of inducing instability. Indeed, if the local controller applied to the plant is always such that it stabilizes the closed loop for a given plant model, it is possible to find situations in which the overall system is unstable due to switching among stable closed-loop systems. The supervisor algorithm must thus be such that it prevents these undesirable situations.

Two types of supervisory algorithms are described hereafter. In the first (gain-scheduling control) it is assumed that there is available a variable whose value indicates the adequate controller to use. In the second (SMMAC) such a variable is not available and the decision on which controller to use relies on the comparison of the observed input/output plant behavior with the one of candidate plant models in a priori defined model bank.

4.1.1 Gain-Scheduling Control

When using gain-scheduling, the active controller is selected at each time instant depending on the value of a (possibly vector) variable known as *scheduling variable*. Scheduling variables define the plant operating regime and can be chosen as the plant input, output, or state, or some auxiliary signals measured in the plant. Figure 4.2 shows an example of a nonlinear plant in which two indicator variables θ_1 and θ_2 define five operating regimes, labeled by the integers 1–5, that correspond to regions, possibly with some overlap, in the space defined by the scheduling variables.

For instance, in a DCSF such as the one used in the simulations depicted in Fig. 1.16 it is reasonable to use either the fluid flow (that is the manipulated variable) or the outlet fluid temperature (that is the process output) as scheduling variable. As discussed in Chap. 2, and in particular in Sect. 2.4.3, both the time scale and the static gain of a linearized DCSF model are directly influenced by the fluid flow at equilibrium. As concluded from Eqs. (2.24) and (2.25), the higher the fluid flow is, the faster the linearized system will be [because the linearized dynamic matrix is proportional to the equilibrium fluid flow, see (2.24)], and the smallest will be the static gain [because the static gain is inversely proportional to the equilibrium fluid flow, as concluded from (2.24) and (2.25)].

One way to design the local controllers for a DCSF is thus to split the range of the possible variations of fluid flow in disjoint intervals between its minimum u_{min}

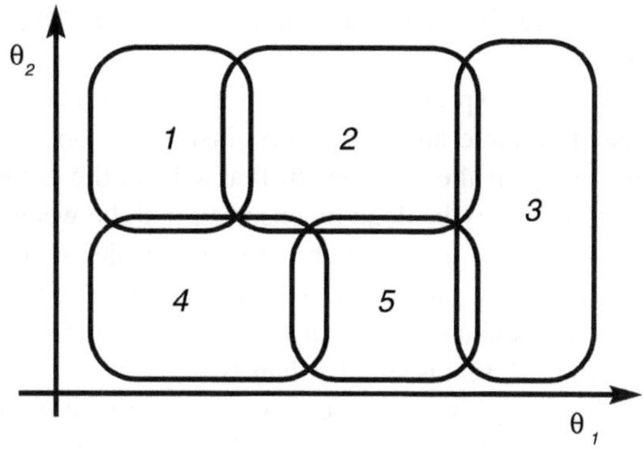

Fig. 4.2 An example of two scheduling variables θ_1 and θ_2 that define five operating regimes, labeled *1–5*

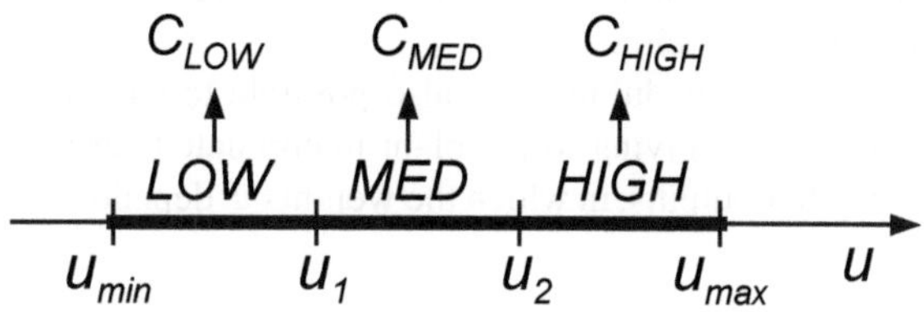

Fig. 4.3 Splitting the range of variation of the fluid flow in a DCSF to use it as a scheduling variable

and the maximum $u_{\max}$ values. For instance, Fig. 4.3 considers three intervals for the flow value that correspond to low (between $u_{\min}$ and u_1), medium (between u_1 and u_2), and high (between u_2 and $u_{\max}$) values of fluid flow and that are labeled, respectively, as LOW, MED, and HIGH. Each of these ranges of the scheduling variable (fluid flow) is associated to a local control, labeled respectively as C_{LOW}, C_{MED} and C_{HIGH}. When the scheduling variable assumes a value inside one of these ranges, the corresponding controller is used. For instance, if the value of the fluid flow u is between $u_{\min}$ and u_1, then C_{LOW} is the controller used.

In a DCSF the outlet fluid temperature can be used as well as an indicator variable since, as explained in Sect. 2.4.3, one is inversely proportional to the other at equilibrium.

Clearly, each local controller has to be designed such as to stabilize the closed loop, yielding the specified performance, when u is assumed to be constant and within the corresponding range of values.

If the scheduling variable changes from one range of values, that define an operating regime, to another, a switching of controllers occurs. In order to prevent a situation in which switching leads to an unstable system, some conditions must be imposed. As shown in Hunt and Johansen (1997), in addition top an adequate design of the local controllers, one must ensure that the plant nonlinearity is sufficiently well approximated by the operating regimes, that the scheduling variable varies sufficiently slow, leading to operating points that change slowly, and that the external noise and disturbances are small. Other type of stability conditions, based on Lyapunov function theory, are obtained in the realm of LPV control theory (Mohammadpour and Scherer 2010).

In order to prevent fast switching, that may cause glitch phenomena in the manipulated variable and, ultimately, instability, one possibility is to include a form of hysteresis in the mechanism that selects the controller from the scheduling variable. This may be done by ensuring that, between two neighbor operating regimes, a transition of a controller A to a controller B happens when the scheduling variable overpasses a value H but, if B is active and the scheduling variable is reduced, there is a switch back from B to A only if the scheduling variable is smaller than a threshold L that is smaller than H.

An alternative is to use a dwell time condition that means that, once a controller is selected, it remains connected to the plant for at least a minimum time interval, known as the *dwell time*. This is the mechanism used in SMMAC to ensure stability, as explained below.

In gain scheduling, it is also possible to smooth transitions between local controllers by applying to the plant manipulated variable a weighted sum of the local controller outputs, in which the weights w depend on the scheduling variable θ. Thus

$$u(t) = \sum_{i=1}^{N} w_i(\theta(t))u_i(t), \tag{4.1}$$

where u_i is the output of local controller C_i, θ is the scheduling variable, w_i is the weight associated to the operating regime i, and t is time. As discussed in Hunt and Johansen (1997), the weights w_i ought to be selected such as to satisfy the following conditions:

1. All the weights are between 0 and 1, $0 \leq w_i(\theta) \leq 1$, for all values of θ. If θ is away from the border of the region that define the operating regime i, then $w_i(\theta) \simeq 1$.
2. For any value of the scheduling variable θ, all the weights add to 1, $\sum_{i=1}^{N} w_i(\theta) = 1$.
3. For any pair of operating regimes indexed by i and j that do not have an overlapping region, then either $w_i = 0$ or $w_j = 0$.

Under these conditions the plant manipulated variable becomes either the output of one of the local controllers, if the scheduling variable is inside and away from the border of one of the operating regimes, or an interpolation of the outputs of the local controllers, if the scheduling variable is close to the border of the corresponding operating regimes that intersect.

4.1.1.1 An Example of Gain-Scheduling

This example illustrates the main features of gain-scheduling in a simple situation that resembles the difficulties found in DCSF control over a wide range of temperature

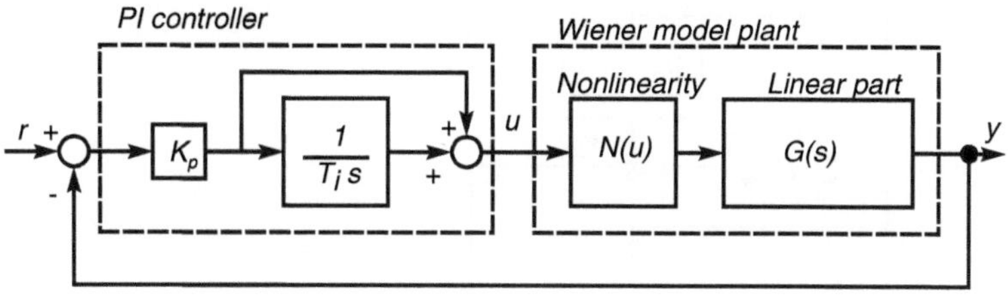

Fig. 4.4 A Wiener-type nonlinear plant controlled with a single PI

Fig. 4.5 Time response of a Wiener-type nonlinear plant controlled with a single PI. There is a degradation in performance when the operating point changes

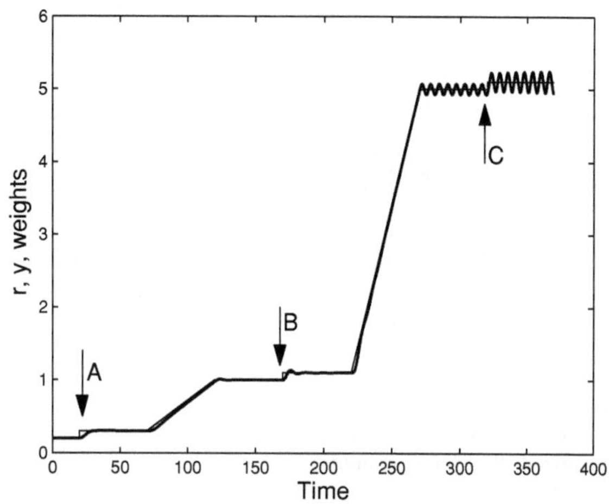

that have been illustrated in Fig. 1.16. Consider a plant with a Wiener model structure that consists of a static nonlinearity in series with a linear dynamic time invariant system. The input u is passed through a nonlinear function $N(.)$ whose output feeds the input of the linear part, that is described by a transfer function $G(s)$ where s is the Laplace variable. Figure 4.4 shows a feedback controller of this plant that uses a constant gain PI controller.

The nonlinear function $N(.)$ is such that it has an incremental gain that increases with its argument. As such, the constant gain controller of Fig. 4.4 experiences difficulties when the operating point changes over a wide range. These difficulties are illustrated in Fig. 4.5 where the reference executes maneuvers in three different operating regions that correspond to values of the output close to 0.2, 1, and 5. Initially, the reference is close to 0.2 and, at the instant indicated by the arrow labeled A, there is a small amplitude step in the reference. If the PI is tuned to a linearized plant dynamics that corresponds to this operating point, then the response of the plant output is adequate.

The reference is then changed, driving the output to a higher value, close to 1. Due to the nonlinearity, the incremental gain is higher and, in response to a small step in the reference (arrow labeled B) there is now some overshoot. If the reference

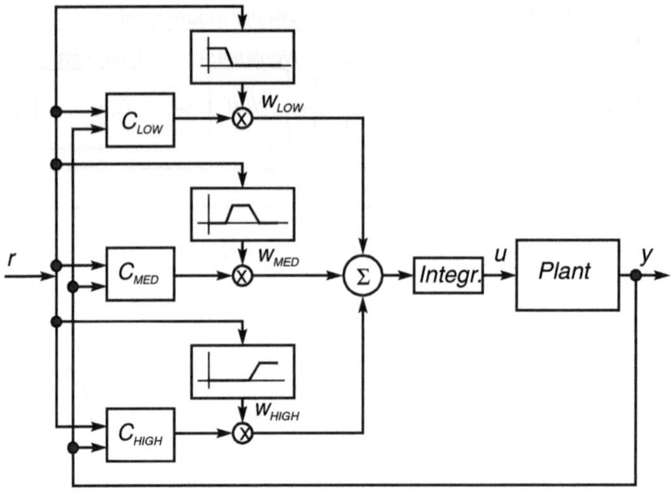

Fig. 4.6 A Wiener-type nonlinear plant controlled with gain-scheduling

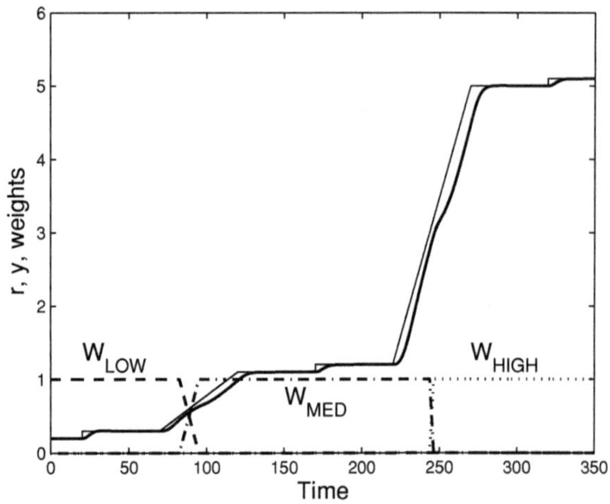

Fig. 4.7 Time response of a Wiener-type nonlinear plant controlled with gain-scheduling. The shape of the step response is now independent of the operating region

is further increased to a value close to 5, the mismatch between the controller tuning and the linearized dynamics is even bigger and the closed-loop becomes unstable (arrow labeled C).

If, instead of relying on a constant gain PI, the gain-scheduling controller of Fig. 4.6 is used, a uniform step response is observed in all the operating regions, as shown in Fig. 4.7.

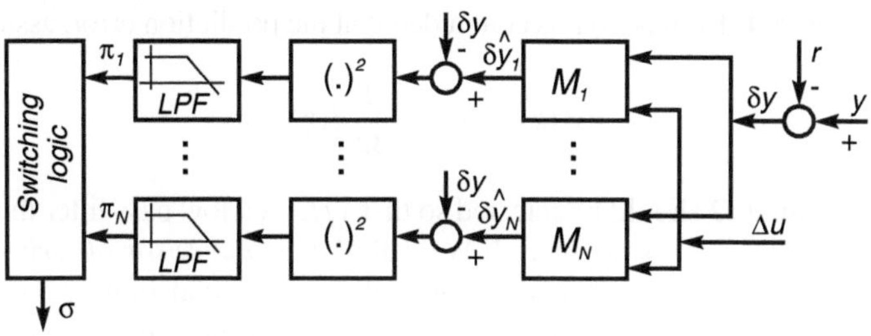

Fig. 4.8 Supervisor for SMMAC

4.1.2 Supervised Multiple Model Adaptive Control

In SMMAC the supervisor makes no assumption about the existence of a scheduling variable that decides which local controller to use but, instead, seeks to take a decision using the structure of Fig. 4.8 that consists of a bank of N models, called *local models*, $\{M_1, \ldots, M_N\}$, whose dynamic behavior is compared with the observed plant input/output signals.

In a discrete time framework, each local model M_i is described by the incremental linear model

$$A_i(q)\delta y(k) = B_i(q)\Delta u(k) + d(k), \tag{4.2}$$

where k is discrete time, $\Delta u(k) = u(k) - u(k-1)$, d is a high frequency disturbance, $\delta y = y - r$ is the deviation of the plant output y with respect to the reference to track r and $A_i(q)$ and $B_i(q)$ are polynomials in the forward shift operator q, with A_i monic (a polynomial is called "monic" if the coefficient of its monomial with biggest degree is 1) and such that the degree of A_i is strictly bigger than the degree of B_i so that the model is causal.

Let $\Omega(q)$ be a Hurwitz polynomial (a polynomial is called "Hurwitz" if all its roots have negative real part) with degree equal to the degree of A_i. The plant model (4.2) may then be written

$$\delta y_i(k) = \frac{\Omega - A_i}{\Omega}\delta y(k) + \frac{B_i}{\Omega}\Delta u(k) + \frac{1}{\Omega}d(k). \tag{4.3}$$

Equation (4.3) the use of predictive models of δy that are of the form

$$\delta\hat{y}_i(k) = \frac{\Omega - A_i}{\Omega}\delta y(k) + \frac{B_i}{\Omega}\Delta u(k), \tag{4.4}$$

where $\delta\hat{y}_i$ denotes the prediction of the plant output (actually, this quantity is the deviation of the plant output with respect to the reference; hereafter we will use this abuse of expression for the sake of simplicity) assuming that the plant dynamics is adequately represented by model M_i.

Subtracting (4.4) from (4.3) it is concluded that the prediction error, assuming M_i correct, is

$$\delta y_i(k) - \delta \hat{y}_i = \frac{1}{\Omega} d(k). \tag{4.5}$$

Hence, polynomial Ω should be selected so that $1/\Omega$ is a low pass filter that washes out the high frequency disturbance d. We call Ω the *supervisor observer polynomial*. The adequate design of this polynomial has a crucial impact on SMMAC performance. In practical situations there is always a mismatch between all the local models and the actual plant that is seen by the models as a disturbance or noise signal. For instance in a DCSF there are high frequency anti-resonance peaks that cannot all be modeled with a finite dimensional model. The supervisor observer polynomial eliminates these high frequency signal components and allows the supervisor to make comparisons between the plant and the local models in the frequency range that is meaningful.

As seen in Fig. 4.8, the supervisor consists of a bank of N parallel predictive models $M_1, \ldots, M_N$ that compute estimates $\delta \hat{y}_i$ of the plant output δy according to (4.4), when assuming that model M_i is true. These estimates are then compared with the actual plant output δy. For this sake, compute first the prediction errors e_i given by

$$e_i(k) = \delta \hat{y}_i(k) - \delta y(k) \tag{4.6}$$

and then the performance indexes π_i that correspond to an estimate of the power of e_i and are updated by the difference equation

$$\pi_i(k) = \lambda_{PI} \pi_i(k-1) + (1 - \lambda_{PI}) e_i^2(k), \tag{4.7}$$

where λ_{PI} is a design parameter to be selected between 0 and 1.

For each M_i, $i = 1, \ldots, N$, a controller C_i is designed. As in gain-scheduling, these controllers are the local controllers shown in Fig. 4.1. In SMMAC, a local controller is selected by the supervisor to be connected to the plant by the switching logic block shown in Fig. 4.8. This block searches for the model M_{i*} with the lowest value of the performance index. If the controller that is currently applied to the plant has been in this state (i. e., connected) for at least τ_D samples, then the controller C_{i*} (that corresponds to the best performance index) is connected to the plant. Otherwise, no switching of controllers occurs.

In other words, according to this algorithm, if a controller is applied to the plant, it remains so for at least τ_D sampling intervals. The parameter τ_D is referred as *dwell time*. In Morse (1996, 1997) it is shown that, if τ_D is greater than or equal to the order of the plant and at least one element of the model bank is a good approximation to the true plant model, then the closed-loop system is stable.

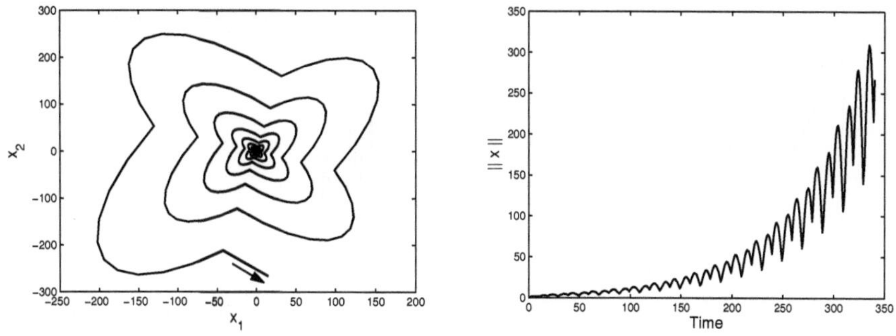

Fig. 4.9 Switching between two stable systems with a dwell time of 10. *Left* A trajectory in the state-space. *Right* The norm of the state as a function of time

4.1.2.1 Example: Switched Systems with a Dwell Time Condition

This example aims at explaining how a time-varying system that results from the switching between two local stable linear systems may be unstable. Furthermore, it also clarifies why it becomes stable if we ensure that each local system, once becoming active, remains so for at least a minimum dwell time.

Consider then the second order time-varying system described by the state-space model

$$x(k+1) = A_{i(k)}x(k), \tag{4.8}$$

where $x \in \mathbb{R}^2$ is the state and $k \in \mathbb{N}$ denotes the discrete time. The index i of the dynamics matrix assumes the values either $i = 1$ or $i = 2$, depending on time k, being

$$A_1 = \begin{bmatrix} 1.17 & -0.22 \\ 0.44 & 0.73 \end{bmatrix} \quad \text{and} \quad A_2 = \begin{bmatrix} 0.73 & -0.22 \\ 0.44 & 1.17 \end{bmatrix}.$$

The index $i(k)$ of the active dynamics matrix A_i is given by

$$i(k) = \begin{cases} 1 & \text{if} \quad 2j\tau_D \le k < (2j+1)\tau_D \\ 2 & \text{if} \quad (2j+1)\tau_D \le k < (2j+2)\tau_D \end{cases}. \tag{4.9}$$

In other words, the sequence of dynamics matrices are A_1 during τ_D time steps, then A_2 during τ_D time steps, with this sequence repeating.

The time-invariant systems corresponding to the situations in which either $i = 1$ or $i = 2$ for all k are both asymptotically stable as can be verified by checking that the eigenvalues of A_1 and A_2 are all strictly inside the unit circle.

There are, however, values of the dwell time τ_D for which the time-varying system becomes unstable. For instance, for $\tau_D = 10$ this system is unstable. As seen in Fig. 4.9, when the time increases there is a lower bound of the state norm that grows without bound. This happens because, for this value of τ_D, the switching between

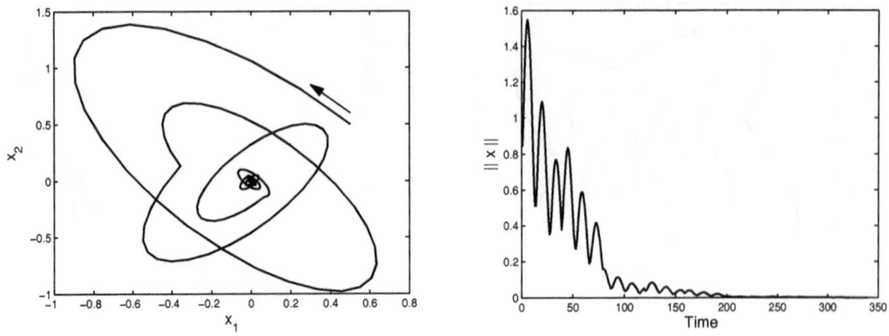

Fig. 4.10 Switching between two stable systems with a dwell time of 40. *Left* A trajectory in the state-space. *Right* The norm of the state as a function of time

local systems always happens in such a way that the state moves away from the origin due to the elongation of the state-space trajectories of the local systems.

Instead, for a large enough value of the dwell time τ_D, switching always happens such that there is an upper-bound of the norm of the cost that decreases to zero. Such a τ_D can always be found because, since both local systems are asymptotically stable, the state vector under each of them will always contract in the long run. For instance, as seen in Fig. 4.10, for $\tau_D = 40$ the time-varying system is stable and the state converges to zero.

Using mathematical arguments it is possible to find a number τ_D^*, that depends on A_1 and A_2, such that, if $\tau \geq \tau_D^*$, then the time-varying system is asymptotically stable.

4.2 Local Controller Design

Local controller design is based on incremental linear models of the form

$$\delta y(k) = \frac{B(q)}{A(q)} \delta u(k), \tag{4.10}$$

where δu and δy are increments of plant input u and output y with respect to the working point and $A(q)$ and $B(q)$ are polynomials in the forward shift operator q, with A monic and $\partial A > \partial B$ so that the local system is causal. As mentioned before, each local system has a different model and is indexed by an integer number between 1 and N. For the sake of simplicity, this index is omitted here both in the system model and in the corresponding controller.

Each local controller is described by

$$R(q)\delta u(k) = -S(q)\delta y(k) + T(q)\delta r(k), \tag{4.11}$$

where $R(q)$, $S(q)$ and $T(q)$ are polynomials in the forward-shift operator q such that $\partial R > \partial S$ and $\partial R > \partial T$ so that the controller is causal and δr is the incremental reference to track. Designing the local controller amounts to compute the coefficients of the polynomials R, S and T such that the specifications are met and the controller has minimal order. For this sake, several methods may be used. One possibility is to use LQG methods. Another is to use MUSMAR (described in Chap. 3) with a plant model and to take for the value of the coefficients of R, S and T the gains at convergence. Still another method is pole placement.

When coupling the plant model (4.10) with the linear controller (4.11) the controlled system is seen to be represented by

$$\delta y(k) = \frac{B(q)T(q)}{A(q)R(q) + B(q)S(q)} \delta r(k), \tag{4.12}$$

when using the pole-placement design procedure, the controller design problem consists of finding polynomials R, S and T such that the controlled system (4.12) has the specified transfer function given by

$$\delta y(k) = \frac{B_m(q)}{A_m(q)} \delta r(k), \tag{4.13}$$

where A_m and B_m are polynomials in the forward shift operator such that $\partial A_m > \partial B_m$, meaning that the specified system is causal, and $\partial A - \partial B \geq \partial A_m - \partial B_m$, meaning that the open loop plant pure delay may not be reduced (a necessary condition to yield a causal controller). The controller polynomials may be computed from the equations

$$T = B'_m A_o, \tag{4.14}$$

$$A(z-1)R + B^- S = A_m A_o, \tag{4.15}$$

where A_o is the observer polynomial, B^- is a factor of B such that $B = B^+ B^-$ with B^+ stable and monic, and B'_m is a factor of the polynomial B_m of the desired closed-loop transfer function, such that $B_m = B^- B'_m$.

4.3 Example: Air Heating Fan

The temperature control of the air heating fan described in Sect. 2.5.4 provides an example to explain the design of a SMMAC system. This example illustrates a general procedure to characterize the way in which the linearized dynamics varies with the working point, the identification of linear models and the design of the local controllers. Important aspects concern the choice of the controller observer polynomial, the design of the local controllers and the design of the supervisor observer

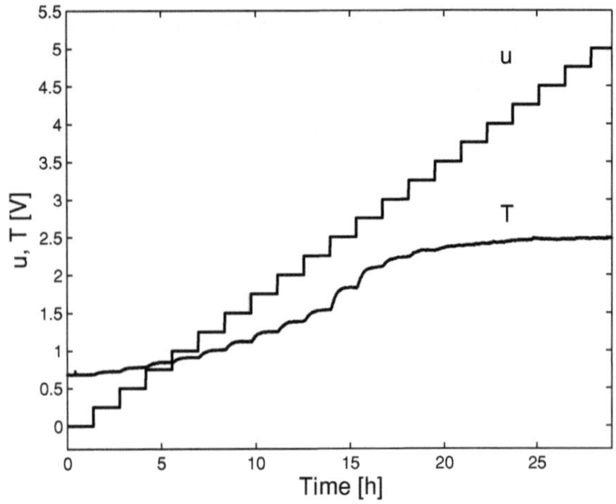

Fig. 4.11 Experiment to estimate the static characteristic of the temperature loop for one value of the air flow

polynomial. The reader may adapt the steps described to design a gain-scheduling or SMMAC controller to a plant of interest to him.

As explained in Sect. 2.5.4, in the plant considered there are two main variables that influence incremental linearized dynamics. These are the average values of air flow and temperature. In the example shown these variables are expressed in electric tension, in which they are transduced, in the range between 0 and 5 V.

In order to assess the nonlinear dependence on the average temperature that defines the working point, the plant has been submitted to an experiment in which the manipulated variable is made to vary according to successive steps, each one taking as initial point the final value of its predecessor. As such, the steps sweep the dynamic and static responses when the manipulated variable ranges from 0 V up to its maximum at 5 V. Each of these experiments has been performed with the flow assuming a constant value. Three sets of experiments have been made, with the flow in what is considered a "low" range, in a "medium" range, and in a "high" range of values. Furthermore, to take into account hysteresis effects, two successive sets of variations have been performed, first with the manipulated variable increasing, and then with the manipulated variable decreasing from step to step.

Figure 4.11 shows the results for one value of the air flow. The approximately triangular signal is the manipulated variable signal used to characterize the dynamics, and the "distorted" signal is the output air temperature. The manipulated variable signal is made to vary as a sequence of step functions, each one starting from the end point of the previous one, and with a duration long enough so that the temperature stabilizes in its new equilibrium point, before the next incremental step is applied.

In this way, around the average value of the incremental amplitude of each input step, one may compute the incremental static gains, shown in Fig. 4.12, and the dominant time constants of the dominant incremental responses, shown in Fig. 4.13. As it is apparent from the above-mentioned plots, the dynamics of the incremental models vary significantly with the average value of the manipulated variable and of

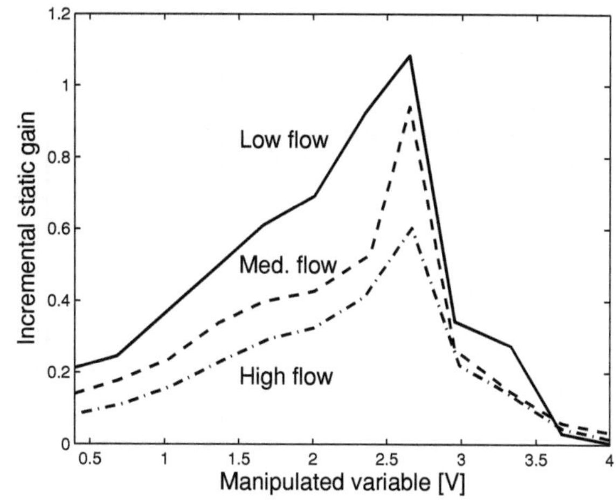

Fig. 4.12 Static gain of incremental temperature models, as a function of the operating point of the manipulated variable, for three different values of the air flow

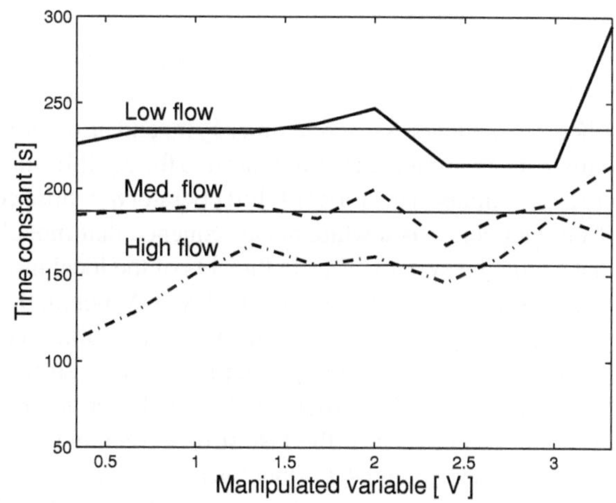

Fig. 4.13 Time constants of incremental models of temperature as a function of the operating point, as defined by the manipulated variable, for three values of the air flow

the air flow. For both low and high values of the manipulated variable the incremental gain is small, whereas for intermediate values it is higher.

The air flow also has a strong influence on the linearized dynamics. As concluded from Fig. 4.13, around the same equilibrium temperature, the system response becomes faster when the air flow increases.

The above results suggest to represent the plant with six local models that correspond to the operating points located in the temperature/air flow space as shown in Fig. 4.14 and that are representative of the main types of dynamical behavior observed in the plant. These local models are identified from plant data obtained from experiments performed around the corresponding operating conditions, and are used as a basis for local controller design.

Each local model has the ARMAX structure

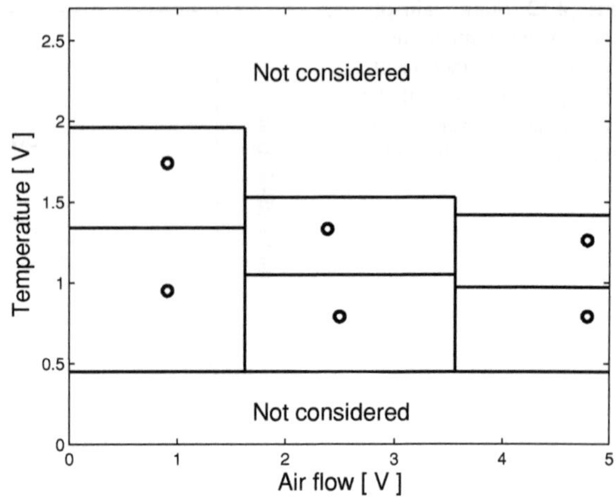

Fig. 4.14 The six local models obtained in the temperature/flow space. Each *dot* represents the operating point for which models have been obtained

$$A(q)\delta y(t) = B(q)\delta u(t) + C(q)e(t), \tag{4.16}$$

where $t \in \mathbb{N}$ denotes discrete time, $\delta y(t)$ and $\delta u(t)$ are the increments of the process output and manipulated variable around the equilibrium point considered, A, B and C are polynomials in the forward shift operator q, whose coefficients are estimated from plant data, and e is a white noise sequence that models the stochastic disturbances that act on the system. The coefficients of the local models and of the corresponding local controllers are listed in the tables of Appendix F.3.

When designing the local controllers according to the pole-placement procedure described in Sect. 4.2, an important issue is the selection of the observer polynomial A_o. In this case, all the roots of A_o are chosen to be real, positive, and equal, and in the range between 0 (the fastest observer) and close to 1 slowest observer). As seen on Figs. 4.15 and 4.17, there is a significant effect of the choice of the roots of the observer polynomial on the manipulated variable. When designing the observer faster the noise influence on the output increases. Furthermore, a trade-off has to be considered in the choice of the observer characteristic polynomial root. Indeed, comparing Figs. 4.15 and 4.16, it is concluded that increasing the observer poles (decreasing the observer bandwidth, as the poles approach 1) reduces the effect of the noise on the manipulated variable (the standard deviation of the manipulated variable decreases, as seen in Fig. 4.15), but also causes a drop on the closed-loop bandwidth (as seen on Fig. 4.16).

An alternative is to choose $A_o = C$, that is optimum in the LQG framework. Figure 4.18 compares two experiments that use an observer polynomial obtained, one by trial and error, and the other by making $A_o = C$.

Finally, Fig. 4.19 shows experimental results when controlling the fan outlet air temperature with SMMAC. This experiment illustrates the capability of the SMMAC controller to yield similar closed-loop responses around two different operating points.

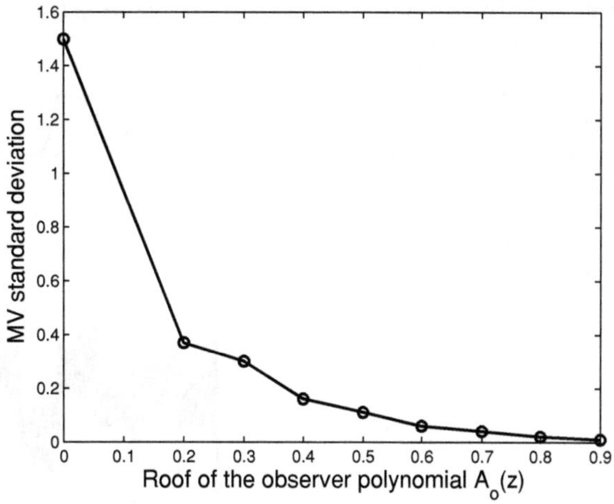

Fig. 4.15 Manipulated variable standard deviation as a function of the modulus of the controller observer polynomial roots

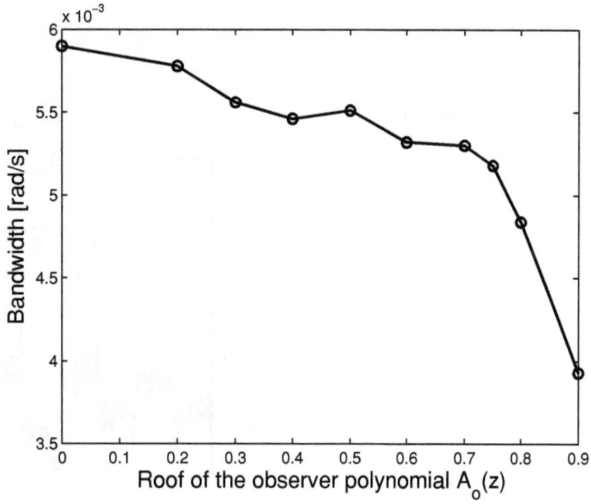

Fig. 4.16 Closed-loop bandwidth as a function of the modulus of the controller observer polynomial roots

4.4 DCSF Application

In order to apply SMMAC to the prototype DCSF considered in this book, the first step consists of characterizing the local models. From the discussion held in Sect. 2.4.3 about Jacobian linearization, it is concluded that the poles of the linearized system depend mainly on the fluid flow (the manipulated variable), while the solar radiation intensity affects only the static gain. This remark leads to consider the local controllers that correspond to the operating conditions described in Table 4.1, where C_{sg} is a constant that adjusts the static gain to reflect different solar radiation characteristics.

To each of the local models, a linear controller, designated "local controller", has been associated by running off-line the MUSMAR algorithm, described in Sect. 3.6,

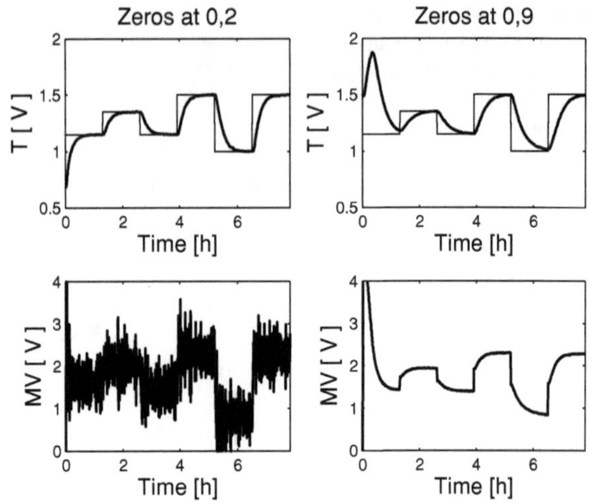

Fig. 4.17 The effect of changing the modulus of the observer eigenvalues: fast observer dynamics (*left*) and slow observer dynamics (*right*). A slower observer leads to a "cleaner" manipulated variable, at the expense of a slight decrease in raise time

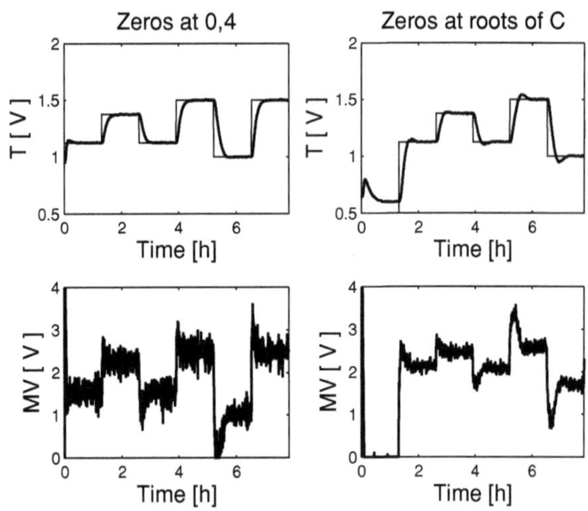

Fig. 4.18 Comparison of manual tuning of the observer eigenvalues (*left*) with the optimal observer eigenvalues (*right*)

applied to the local model, and taking as local controllers the ones obtained after convergence is achieved. In the experiments described hereafter, the sampling interval is 30 s and the dwell time is $T_D = 10$ s. The prediction horizon used for MUSMAR is $T = 10$ samples. The value of the penalty ρ on the manipulated variable depends on the local controller and is shown in Table 4.1.

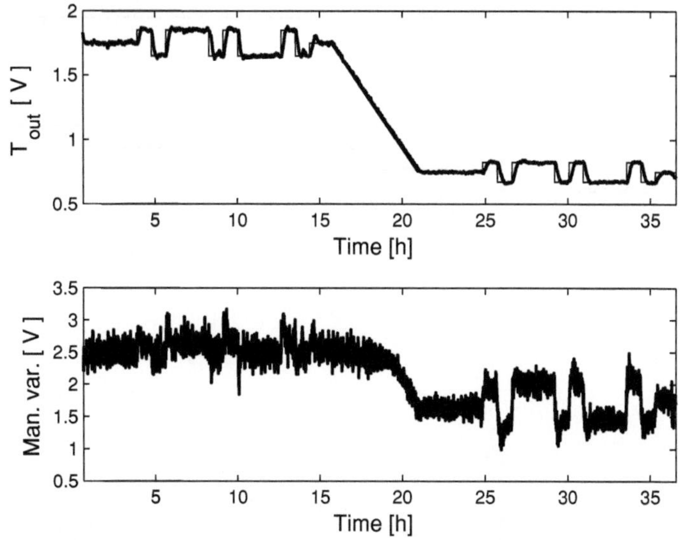

Fig. 4.19 Experimental results when controlling the outlet air temperature with SMMAC in the air heating fan plant

Table 4.1 Local controllers of the DCSF considered in different operating points

Controller n.	Flow (L/s)	C_{sg}	ρ
0	5	1.00	100
1	5	1.50	100
2	5	0.66	50
3	5	0.33	10
4	7	1.00	70
5	7	1.50	150
6	7	1.50	150
7	7	1.0	100
8	3.5	1.0	1000

4.4.1 DCSF Control with MMAC: Experiment 1

In experiment 1, Figs. 4.20 and 4.21 show experimental results obtained with MMAC in the DCSF considered. Figure 4.20 depicts the time evolution of the main variables of the control loop, namely the manipulated variable (fluid flow in (l/s), multiplied by 10 for graphical convenience), the reference, with a decreasing step followed by a sequence of three increasing steps, and the outlet fluid flow, that closely follows the reference. Figure 4.21 shows the index of the active local controller as a function of time. This index changes not only after reference changes, but also due to radiation changes that may happen during periods in which the reference is kept constant.

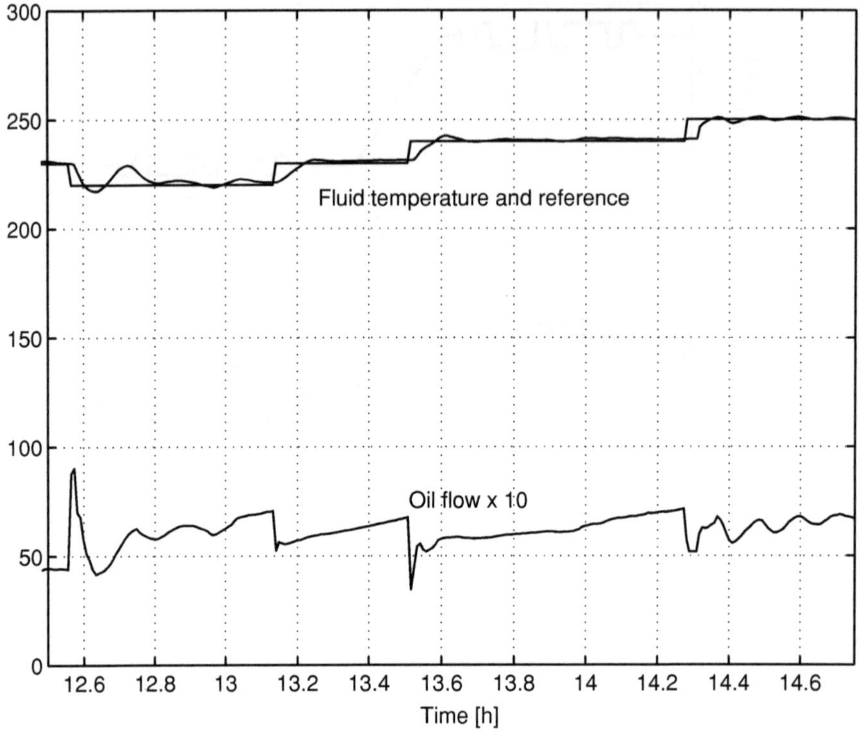

Fig. 4.20 Controlling a DCSF with SMMAC: experiment 1. Reference, output temperature and manipulated variable (°C) [fluid flow in (l/s) multiplied by 10]. Time in (h)

From the analysis of Fig. 4.20 it is concluded that there is a small overshoot and a settling time of about 6 min, that is considered to be very good in this plant.

4.4.2 DCSF Control with MMAC: Experiment 2

Experiment 2 illustrates the adaptation capability of the SMMAC algorithm in responses to changes in either the value of incident solar radiation or the temperature reference. Figures 4.22 and 4.23 refer to experiment 2 on the use of SMMAC on the DCSF. Figure 4.22 shows the reference and the outlet fluid temperature (above) and the fluid flow (manipulated variable) and the index of the active controller as a function of time (below). Figure 4.23 represents the temporal evolution of the solar radiation during the experiment.

In experiment 2, when the temperature suddenly drops due to a passing cloud, the active controller changes from 7 to 5. The reference increase causes a decrease in the fluid flow and forces a further change of the active controller, that becomes

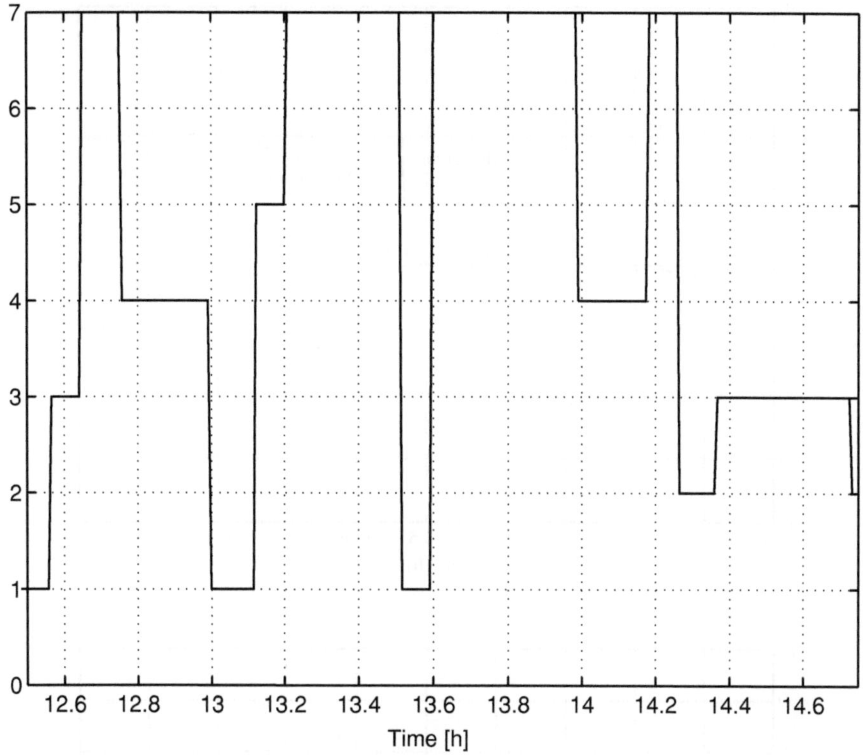

Fig. 4.21 Controlling a DCSF with SMMAC: experiment 1. Index of the active controller (supervisor output). Time in (h)

controller number 2. After the radiation recovers its initial value and the temperature settles down, the active controller index returns to 7.

4.4.3 DCSF Control with MMAC: Experiment 3

In experiments 1 and 2, the index of the active local controller is often 7. Experiment 3 answers the question about what happens when the number of local controllers is drastically reduced, by keeping only controller number 7, designed for a fluid flow of 7 l/s, that is close to the maximum flow supported by the plant, and controller number 8, that is designed to match the much smaller flow of 3, 5 l/s. From the results in Fig. 4.24, a significant performance degradation is apparent, with both the output temperature and the fluid flow presenting undesirable oscillations. This example illustrates, therefore, the importance of selecting the grid of local controllers so as to cover adequately the possible plant dynamics outcomes.

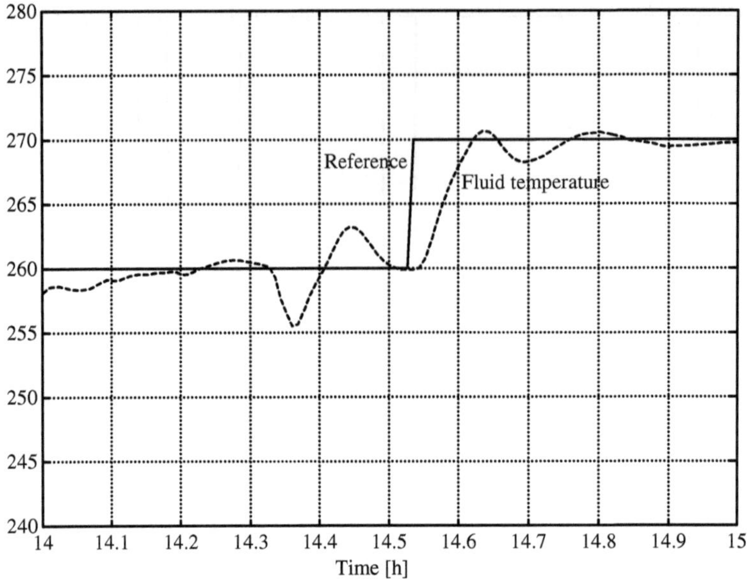

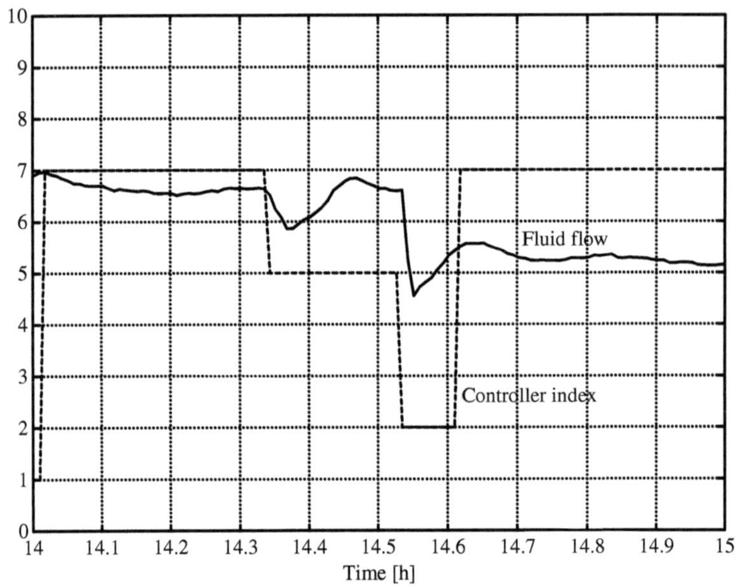

Fig. 4.22 Controlling a DCSF with SMMAC: experiment 2. Reference and output temperature (°C) (*above*), and index of active controller (*dashed line*) and fluid flow in (l/s) (*below*). Time in (h)

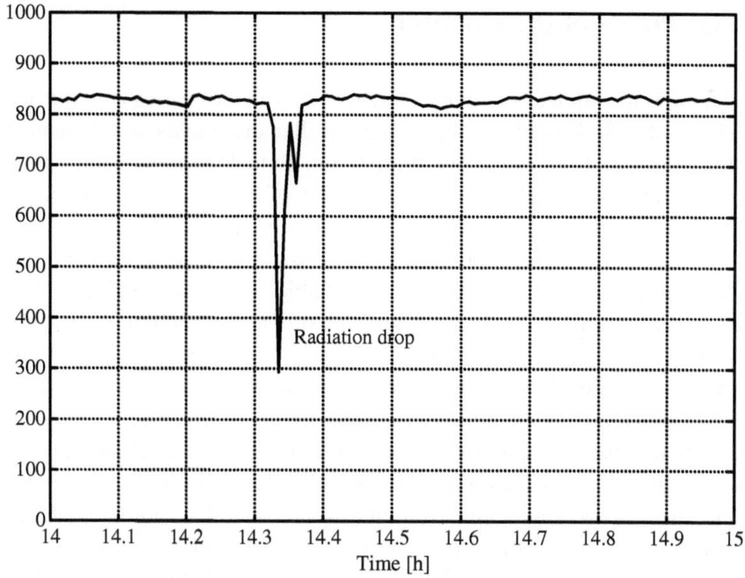

Fig. 4.23 Controlling a DCSF with SMMAC: experiment 2. Solar radiation (W/s^2). Time in (h)

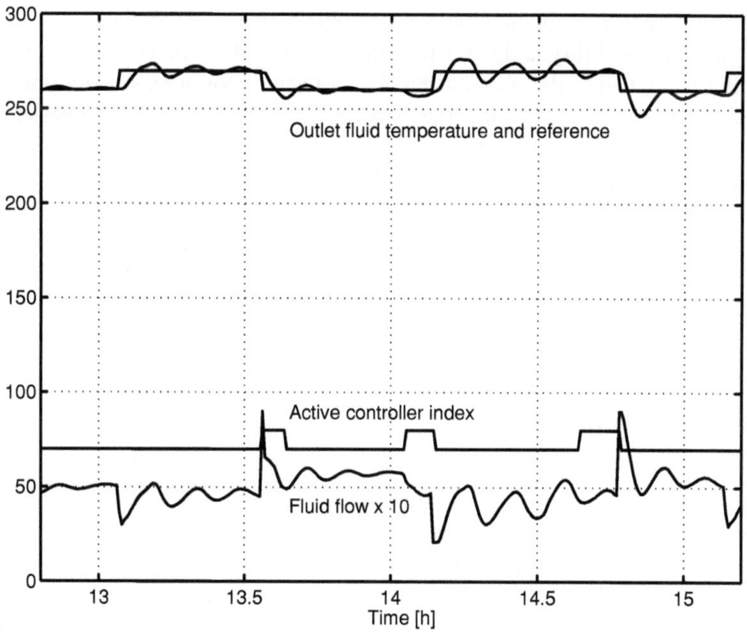

Fig. 4.24 Controlling a DCSF with SMMAC: experiment 3. Reference and output temperature (°C), fluid flow (l/s) multiplied by 10 and the index of the active local controller multiplied by 10. Time in (h)

4.5 Main Points of the Chapter

Since a DCSF is a nonlinear system, its linearized dynamics varies with the operating point. This makes it difficult to design a controller that yields an adequate performance for large variations of the operating point. To tackle this problem this chapter proposes an approach in which a finite number of linear controllers, that form a model bank, is patched so as to obtain a suitable nonlinear controller.

In a first approach, this can be done with classical gain-scheduling in which the controller that generates the manipulated variable is selected based on the value assumed by a so-called scheduling variable that provides a good indication about the plant dynamics around the equilibrium point being considered. In a DCSF, either the fluid flow or the outlet fluid temperature can be used as scheduling variables since they impose the dominant time-scale and static gain.

When an adequate scheduling variable is not available SMMAC proposes an approache in which the selection of the controller that becomes active is made by comparing the observed plant input/output behavior with the simulated input/output behavior of a set of models in a priori defined model bank. To each model in this bank a controller is associated such that, if the model is controlled with the model, the closed-loop is stable and the desired specifications are met. The model that yields the closest match is then used to select the corresponding controller whose output is a candidate to generate the manipulated variable to apply to the plant.

To guarantee stability, a dwell time condition is enforced. This means that, if a controller is applied to the plant, it remains connected for at least a minimum time interval, known as the dwell time. Furthermore, in order to ensure bumpless transfer when switching, controllers generate increments of the manipulated variable and feed a common integrator that generates the manipulated variable actually fed to the plant.

Important issues are the design of the local controllers, the choice of the observer polynomials, and the number of local models that must "cover" the possibilities of plant dynamics outcomes.

4.6 Bibliographic Notes

Gain-scheduling started to be first developed in relation to practical engineering applications, with very little theory supporting it (Rugh 1991). The first work to address the so-called classical approach to gain-scheduling using a systematic theoretical basis was Shamma (1988). See also Shamma and Athans (1990). With the introduction of LPV systems (Shamma and Athans 1991), controller design was formulated as a convex optimization problem with LMI constraints thereby avoiding the interpolation drawbacks of the classical approach (Apkarian et al. 1995; Balas 2002; Mohammadpour and Scherer 2010).

There is a rich literature on applications of gain-scheduling and LPV design methods to a wide variety of engineering systems, including several works that use the classical approach to DCSF and solar furnace control. In Camacho et al. (1994a,b), Pickardt (2000) adaptation in Generalized Predictive Control (GPC) is achieved by controller interpolation using a gain-scheduling approach, whereas Johansen et al. (2000) uses pole placement to design the local controllers. Classical gain-scheduling has also successfully been applied to solar furnaces (Berenguel et al. 1999).

An early version of SMMAC was proposed in Lainiotis (1976). Progress in the understanding of robustness and other issues related to computer control lead to a renewed interest after the works (Morse 1996, 1997; Narendra and Balakrishnan 1997) that result in many works of which (Zhivoglyadov et al. 2001; Anderson et al. 2000, 2001; Angeli and Mosca 2002) are significant samples.

The experimental application to a DCSF of approach to SMMAC that uses MUSMAR at convergence to design local controllers is described in Rato et al. (1997).

References

Anderson B, Brinsmead T, De Bruyne F, Hespanha J, Liberzon D, Morse AS (2000) Multiple model adaptive control. Part 1: finite controller coverings. Int J Robust Nonlinear Control 10:909–929

Anderson B, Brinsmead T, De Bruyne F, Hespanha J, Liberzon D, Morse AS (2001) Multiple model adaptive control. Part 2: switching. Int J Robust Nonlinear Control 11:479–496

Angeli D, Mosca E (2002) Lyapunov-based switching supervising control of nonlinear uncertain systems. IEEE Trans Autom Control 47:500–505

Apkarian P, Gahinet P, Becker G (1995) Self-scheduled H_∞ control of linear parameter varying systems: a design example. Automatica 31(9):1251–1261

Balas GJ (2002) Linear, parameter-varying control and its application to a turbofan engine. Int J Robust Nonlinear Control 12:763–796

Berenguel M, Camacho EF, García-Martín FJ, Rubio FR (1999) Temperature control of a solar furnace. IEEE Control Syst 19(1):8–24

Camacho EF, Berenguel M, Bordóns C (1994a) Adaptive generalized predictive control of a distributed collector field. IEEE Trans Control Syst Tech 2(4):462–467

Camacho EF, Berenguel M, Rubio F (1994b) Application of a gain scheduling generalized predictive controller to a solar power plant. Control Eng Pract 2(2):227–238

Hanus R, Kinnaert M, Henrotte JL (1987) Conditioning technique, a general anti-windup and bumpless transfer method. Automatica 23(6):729–739

Hunt KJ, Johansen TA (1997) Design and analysis of gain-scheduled control using local controller networks. Int J Control 66(5):619–651

Johansen TA, Hunt K, Petersen I (2000) Gain-scheduled control of a solar power plant. Control Eng Pract 8(9):1011–1022

Lainiotis GG (1976) Partitioning: a unifying framework for adaptive systems, II: control. IEEE Proc 64(8):1182–1198

Lourenço JM, Lemos JM (2008) Predictive adaptive control of plants with online structural changes based on multiple models. Int J Adapt Control Signal Process 22(8):774–794

Lourenço JM, Lemos JM (2006) Learning in switching multiple model adaptive control. IEEE Instrum Meas Mag 9(3):24–29

Mohammadpour J, Scherer CW (eds) (2010) Control of linear parameter varying systems with applications. Springer, Heidelberg

Morse S (1996) Supervisory control of families of linear set-point controllers—part 1: exact matching. IEEE Trans Autom Control 41(10):1413–1431
Morse S (1997) Supervisory control of families of linear set-point controllers—part 2: robustness. IEEE Trans Autom Control 42:1500–1515
Narendra K, Balakrishnan J (1997) Adaptive control using multiple models. IEEE Trans Autom Control 42(2):171–187
Pickardt R (2000) Adaptive control of a solar power plant using a multi-model control. IEE Proc Theory Appl 147(5):493–500
Rato LM, Borrelli D, Mosca E, Lemos JM, Balsa P (1997) MUSMAR based switching control of a solar collector field. In: Proceedings of the European control conference, ECC97, Brussells, Belgium
Rugh WJ (1991) Analytical framework of gain scheduling. IEEE Control Syst Mag 11(1):79–84
Shamma JS (1988) Analysis and design of gain-scheduled control systems. PhD Thesis, Department of Mechanical Engineering, Massachusets Institute of Technology
Shamma JS, Athans M (1990) Analysis of gain scheduled control for nonlinear plants. IEEE Trans Autom Control 35:898–907
Shamma JS, Athans M (1991) Guaranteed properties of gain scheduled control for linear parameter varying plants. Automatica 27:559–564
Turner MC, Walker DJ (2000) Linear quadratic bumpless transfer. Automatica 36:1089–1101
Zaccarian L, Teel AR (2002) A common framework for anti-windup, bumpless transfer and reliable designs. Automatica 38:1734–1744
Zaccarian L, Teel AR (2005) The $\mathcal{L}_2(l_2)$ bumpless transfer problem for linear plants: its definition and solution. Automatica 41:1273–1280
Zhivoglyadov PV, Middleton RH, Fu M (2001) Further results on localization-based switching adaptive control. Automatica 37:257–263

Chapter 5
Nonlinear Adaptive Predictive Control

The embedding of nonlinear models in predictive control yields the capacity of operating the plant inside a large domain while achieving a constant performance. Indeed, if linear models are used to design the controller, when the reference moves away from the nominal working point, predictions become poor and the controller performance degrades. Adaptive algorithms such as the ones described in Chap. 3 can counteract this degradation only up to a certain point, since they are able to adjust controller gains tuning them for the new operating point, but they are not able to cope with the fast transient associated to large references excursions. This motivated techniques such as nonlinear MPC (Grüne and Pannek 2011).

As already stated, a general presentation of the methods of nonlinear MPC is outside the scope of this book. Instead, we focus on the class of methods that have a direct relation with adaptive control of DCSFs. Hence, in this chapter a nonlinear MPC algorithm for DCSFs is obtained by taking advantage of the model distributed parameter model given by the PDE (2.5) studied in Chap. 2 and repeated here for the sake of clarity

$$\frac{\partial}{\partial t}T(z, t) = -u\frac{\partial}{\partial z}T(z, t) + \alpha R(t) - \gamma T(z, t), \tag{5.1}$$

As discussed in Chap. 2 this system has a bilinear structure since the manipulated variable u multiplies the state. By making a change in the manipulated variable, together with a change in the time variable, implemented through a time varying sampling interval, it is possible to obtain exactly linear plant models, either input/output or state-space models. Figure 5.1 shows a simplified block diagram of the controller structure that results from this approach. Using discrete models that are obtained from a time varying sampler the controller computes a virtual manipulated variable by minimizing a quadratic cost in a receding horizon sense. From this virtual control, and from the solar radiation power, the actual flow to use in the DCSF as well as the sampling interval to be used by the clock are computed.

The details of the above procedure are explained hereafter.

J. M. Lemos et al., *Adaptive Control of Solar Energy Collector Systems*, Advances in Industrial Control, DOI: 10.1007/978-3-319-06853-4_5,

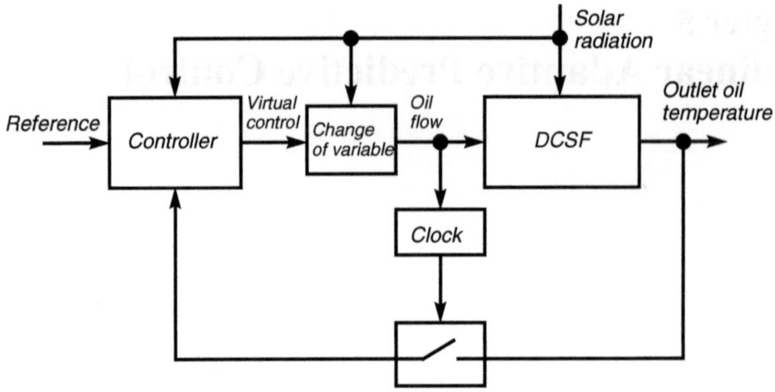

Fig. 5.1 Structure of nonlinear MPC with time varying sampling

5.1 Warped Time Output Prediction Models

As shown in Sect. 2.3, the solution of (5.1) can be approximated by (2.13) that, in terms of the fluid flow F and assuming $\gamma = 0$, is written for a generic Δ as

$$T(z, t + \Delta) = T\left(z - \frac{1}{A_f} F(t)\Delta,\ t\right) + \alpha R(t)\Delta. \tag{5.2}$$

In order to motivate the use of a variable sampling time period, recall Sect. 2.2.3. In the (idealized) situation in which there is neither heating by the sun ($\alpha = 0$) nor losses ($\gamma = 0$) the temperature distribution in the pipe evolves in time such that it is constant along the characteristic lines given by the locus in the plane $[z,\ t]$ defined by

$$z = z_o + \frac{1}{A_f} \int_0^t F(\sigma) d\sigma, \tag{5.3}$$

where z_0 is any value in the interval $[0,\ L]$. If the fluid flow is constant, the characteristic curves are straight lines. Consider the "warped" timescale τ related with "physical" time t by

$$\tau \triangleq \frac{1}{F_0} \int_0^t F(\sigma) d\sigma, \tag{5.4}$$

where F_0 is a constant with dimensions of flow or, equivalently, by

$$\frac{d\tau}{dt} = \frac{F(t)}{F_0}. \tag{5.5}$$

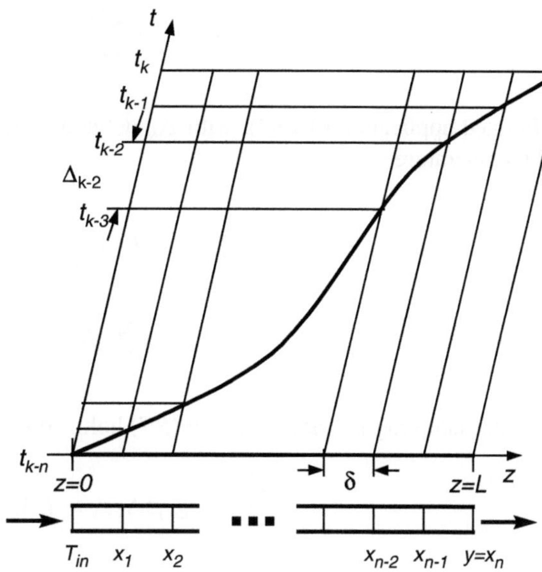

Fig. 5.2 Example of a characteristic line showing that constant sampling in the space variable z implies varying sampling in the time variable t

Let the active part of the pipe, that is to say, the part between $z = 0$ and $z = L$ be divided in n segments of length δ. As shown in Fig. 5.2 the fluid element in the first segment will exactly exit the pipe after n time sampling periods only if a varying sampling interval in time is used. Since in the timescale τ the characteristic curves become straight lines and samples equally spaced in space correspond to samples equally spaced in time, it is natural to consider time scaling in this warped scale. Therefore, considering (5.3), the time sampling interval Δ_k (that may vary with the discrete-time epoch k) given by

$$\Delta_k = t_{k+1} - t_k \tag{5.6}$$

is computed from

$$\Delta_k = \frac{A_f \delta}{F(t_k)}, \tag{5.7}$$

where it is assumed that the fluid flow is kept constant in the interval $[t_k, t_{k+1}]$. Equation (5.7) means that the sampling interval is chosen such that, during it, the fluid in each pipe volume of length δ is transferred to the following volume element. Actually, (5.7) and (5.7) are a way of implementing an approximation to the transformed timescale defined by (5.4). With the sampling defined by (5.7) the fluid contained in the first volume element at time t_{k-n} will exit the pipe exactly after n samples, at the continuous time instant given by

$$t_k = t_{k-n} + \sum_{i=1}^{n} \Delta_{k-i}. \tag{5.8}$$

Iterated application of (5.2), with Δ_k defined by (5.6, 5.7) yields for the outlet fluid temperature

$$T\left(L,\, t_{k-n} + \sum_{i=1}^{n} \Delta_{k-i}\right) = T\left(L - \frac{1}{A_f}\sum_{i=1}^{n} F(t_{k-i})\Delta_{k-i},\, t_{k-n}\right) \tag{5.9}$$
$$+\alpha \sum_{i=1}^{n} R(t_{k-i})\Delta_{k-i}.$$

Since the sampling is made such that (5.7) holds, it follows that

$$\sum_{i=1}^{n} F(t_{k-i})\Delta_{k-i} = LA_f \tag{5.10}$$

and, considering (5.8), (5.9) becomes

$$T(L,\, t_k) = T(0,\, t_{k-n}) + \alpha \sum_{i=1}^{n} R(t_{k-i})\Delta_{k-i}. \tag{5.11}$$

Let

$$T_{\text{in}}(k-n) \triangleq T(0,\, t_{k-n})$$
$$T_{\text{out}}(k) \triangleq T(L,\, t_k)$$

denote the fluid temperature, respectively, at the input and at the output of the active part of the pipe and take as manipulated variable u the inverse of the flow multiplied by $A_f\delta$

$$u(k) \triangleq \frac{A_f\delta}{F(t_k)}. \tag{5.12}$$

From the above choices and the fact that the sampling interval verifies (5.7), it is concluded that model (5.11) can be written as

$$T_{\text{out}}(k) = \alpha \sum_{i=1}^{n} R(k-i)u(k-i) + \beta T_{\text{in}}(k-n). \tag{5.13}$$

Since only discrete times are considered, for simplicity of notation $R(k-i)$ denotes $R(t_{k-i})$. The parameter β has been inserted to take into account losses.

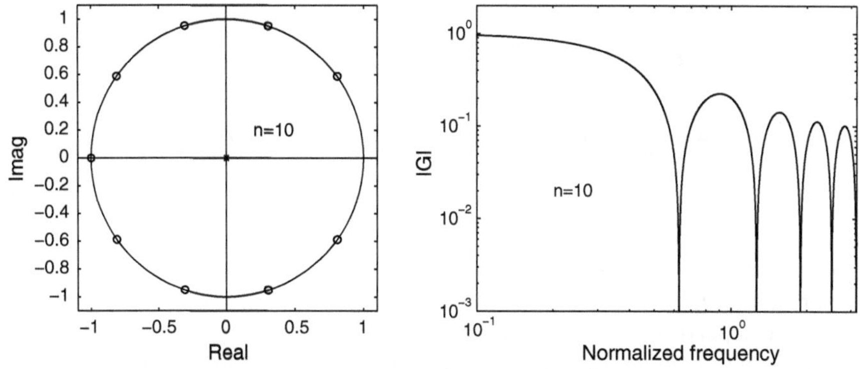

Fig. 5.3 Pole/zero map and frequency response of the warped time DCSF model, with the active part of the pipe divided into $n = 10$ segments, showing anti-resonant peaks

Equation (5.13) provides a predictive model and will be used as a basis for nonlinear predictive control design. It is remarked that, assuming no losses ($\beta = 1$) and a constant radiation R, Eq. (5.13) defines a finite-impulse response (FIR) linear model. For illustrative purposes, Fig. 5.3 shows the pole-zero map of the resulting transfer function and the corresponding frequency response, when the number of elements in which the pipe is divided is $n = 10$. As seen, there are anti-resonant peaks due to the zeros over the unit circle. Although for a different situation (because here the linear model stems in an exact way from a "warped" timescale and not from Jacobian linearization) this is in relation with the results of Berenguel and Camacho (1996), Meaburn and Hughes (1994). A relation may also be established with Chap. 6 in which the internal dynamics resulting from feedback linearization is shown to be oscillatory.

5.2 Output-Based Predictive Control

The control law to consider in this chapter is designed so as to minimize in a receding horizon sense the quadratic cost deterministic functional given by

$$J_k = \sum_{j=1}^{T} \left[T_{out}^*(k+j) - T_{out}(k+j)\right]^2, \tag{5.14}$$

where T (integer) is the prediction horizon and T_{out}^* is the reference to track.

From (5.13) it is seen that the outlet fluid temperature along the prediction horizon verifies the following predictive model:

$$T_{\text{out}}(k+i) = \alpha \sum_{j=1}^{i} R(k+j-1)u(k+j-1) + \alpha \sum_{j=i+1}^{n} R(k+j-1)u(k+i-j) + \beta T_{\text{in}}(k+i-n), \quad i = 1, \ldots, T. \tag{5.15}$$

The prediction of the outlet fluid temperature depends on the future values of the solar radiation R and of the manipulated variable u. If the horizon is such that $T \leq n$ there is no dependency on the future values of the inlet fluid temperature T_{in} because, with the time sampling used, an element of fluid that enters the active part needs n discrete time periods to reach the output.

It is shown in Appendix F.1 that, by inserting (5.15) in the cost (5.14) and minimizing J_k with respect to $u(k)$, assuming that u and the solar radiation R are constants along the prediction horizon, results in

$$u(k) = \frac{1}{\alpha R(k) \sum_{i=1}^{T} i^2} \left[\sum_{i=1}^{T} i \left(T_{\text{out}}^*(k+i) - \beta T_{\text{in}}(k-n+i) \right) - \alpha \sum_{i=1}^{T} i \times \sum_{j=i+1}^{n} R(k+i-j)u(k+i-j) \right] \tag{5.16}$$

The minimization of (5.14) is performed using a receding horizon strategy according to which, of the whole sequence of samples of u defined along the prediction horizon that minimize J_k, only the first, given by (5.16) is actually applied to the plant input at time k, the whole optimization procedure being repeated in the next sampling interval.

The following WARTIC-*i/o* (WARped TIme Controller-based on *i/o* model) algorithm (Silva et al. 2003a) summarizes the control strategy just described:

WARTIC-*i/o* algorithm.
At the beginning of each sampling interval, recursively execute the following steps:

1. On the basis of plant measurements, estimate by RLS the parameters α and β in the model (5.13).
2. Compute the virtual manipulated variable $u(k)$ by (5.16).
3. Compute the fluid flow to apply to the plant by (5.12).
4. Compute the duration of the sampling interval during which the flow obtained in the previous step is applied to the plant by

$$\Delta_k = u(k). \tag{5.17}$$

□

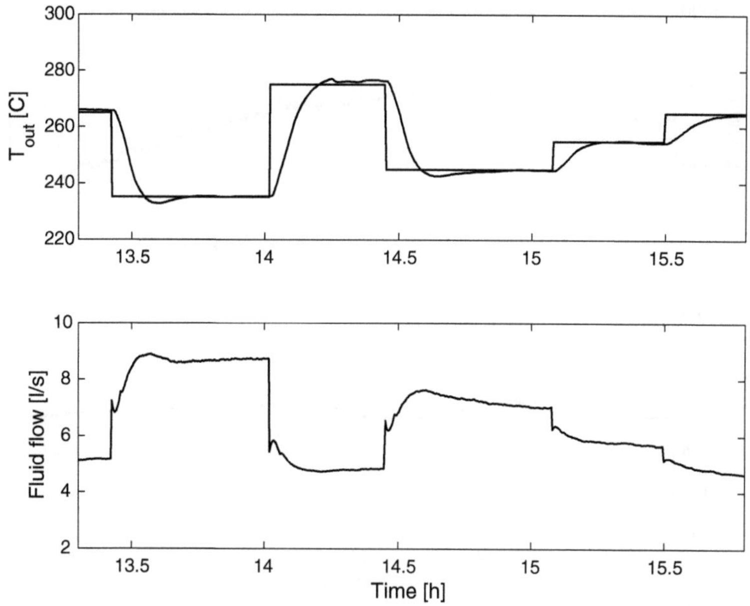

Fig. 5.4 Experimental results with WARTIC-*i/o*. Fluid outlet temperature and reference (*above*) and fluid flow (*below*)

This algorithm is mainly a feedforward algorithm that compensates by anticipation the effect of the two main accessible disturbances that affect the DCSF: Solar radiation and inlet fluid temperature. As such, WARTIC-*i/o* has the advantages of feedforward (fast response), but the drawback of having feedback only through the estimation of parameters α and β. The algorithm WARTIC-*state*, to be explained in the next section, overcomes this inconvenient.

*5.2.1 Experimental Results with WARTIC-*i/o

Figures 5.4 through 5.6 show an example of applying WARTIC-*i/o* to the DCSF used as a prototype for experiment . The controller is configured using $T = 9$ and $n = 10$. Figure 5.4 shows the fluid outlet temperature superimposed on the reference (above) and the fluid flow (below). It is remarked that fluid flow is an image of the sampling period that varies in time according to (5.7). Figure 5.5 shows the accessible disturbances, respectively, the solar radiation power (above) and the inlet fluid temperature (below). The inlet fluid temperature is approximately constant, while solar radiation power changes according to daytime, with fast fluctuations due to scattered atmospheric moisture, but without fast and strong changes due to passing clouds. Finally, Fig. 5.6 shows the evolution of the estimates of the coefficients α and

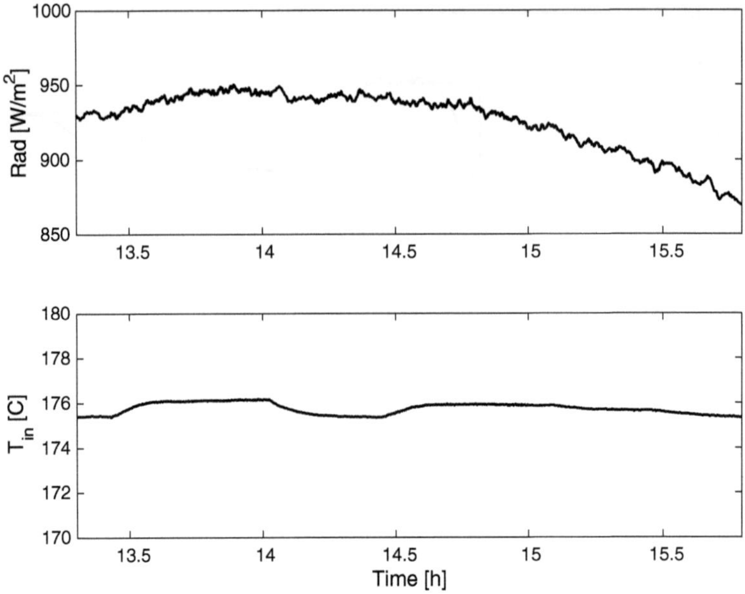

Fig. 5.5 Experimental results with WARTIC-*i/o*. Solar radiation (*above*) and fluid inlet temperature (*below*)

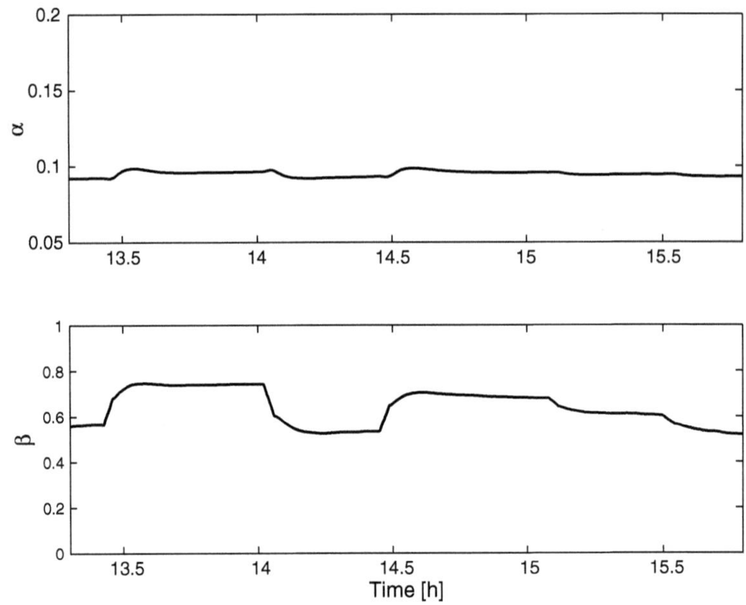

Fig. 5.6 Experimental results with WARTIC-*i/o*. Parameter estimates

β of the time warped-discrete model (5.13). While α is approximately constant, with changes less than 10 % of its nominal value, β presents larger variations that are close to the sampling interval (compare the graphic for β with the graphic for the flow that is proportional to the sampling time), a fact that can be anticipated by relating β with γ.

The most significant issue illustrated by this experiment is that it is possible to apply jumps of high value to the reference and obtain a fast response without overshoot. In Fig. 5.4 a jump of 40 °C with no overshoot and a settling time of about 12 min is observed. This type of performance is typical of control design methods that explicitly take into consideration the nonlinear distributed dynamics of DCSFs. For instance in Johansen and Storaa (2002) experimental results including a jump of 30 °C are reported.

5.3 State-Based Predictive Control

An adaptive nonlinear MPC control algorithm based on state-space models is now explained. Again nonuniform sampling depending on the fluid flow allows to obtain an exactly linear state-space model in discrete time.

5.3.1 Linear State-Space Model Based on Nonuniform Sampling

In order to obtain a nonlinear MPC algorithm based on state feedback, the state that consists of the temperature at the end of each of the n elements of the pipe is considered. More specifically the state variables are defined by

$$x_i(t) \triangleq T(i\delta, t) \tag{5.18}$$

for $i = 1, \ldots, n$. This is the state defined in Sect. 2.4 and corresponds to the nodes of the grid shown in Fig. 2.8. The difference in the state-space model considered here to the one in Sect. 2.4 is that, here, we take advantage of the time varying sampling defined by (5.7) in order to obtain a linear discrete time state-space model. It is stressed that linearity is achieved not by any form of Jacobian approximation but by a nonlinear transform applied to the model.

By using similar arguments as employed to deduce (5.13) and considering the sampling defined by (5.7) it is possible to show that the different state variables are related by

$$x_j(k+1) = \beta x_{j-1}(k) + \alpha R_{av}(k)\Delta_k, \tag{5.19}$$

where $R_{av}(k)$ is the average solar radiation power received during the sampling interval between t_k and t_{k+1}, given by

$$R_{av}(k) = \frac{1}{\Delta_k} \int_{t_k}^{t_{k+1}} R(\sigma) d\sigma. \tag{5.20}$$

Using (5.7), Eq. (5.19) can be written as

$$x_j(k+1) = \beta x_{j-1}(k) + \alpha A_f \delta \frac{R_{av(k)}}{F(k)}, \quad j = 1, \ldots, n. \tag{5.21}$$

Furthermore, define the output

$$y(k) \stackrel{\Delta}{=} x_n(k) = T(L, t_k), \tag{5.22}$$

the accessible disturbance input given by the inlet fluid temperature

$$d(k) \stackrel{\Delta}{=} x_0(k) = T(0, t_k) \tag{5.23}$$

and the virtual manipulated variable

$$u(k) = \alpha A_f \delta \frac{R_{av}(k)}{F(k)}. \tag{5.24}$$

The state vector $x = \begin{bmatrix} x_1(k) & \ldots & x_n(k) \end{bmatrix}^T$ satisfies thus the linear model with dynamical equation

$$x(k+1) = Ax(k) + Bu(k) + Dd(k) \tag{5.25}$$

and output equation

$$y(k) = Cx(k), \tag{5.26}$$

where

$$A = \begin{bmatrix} 0 & \ldots & 0 \\ & \beta I_{n-1} & \vdots \\ & & 0 \end{bmatrix}, \quad B = \begin{bmatrix} 1 \\ 1 \\ \vdots \\ 1 \end{bmatrix}, \quad D = \begin{bmatrix} 1 \\ 0 \\ \vdots \\ 0 \end{bmatrix},$$
$$C = \begin{bmatrix} 0 & \ldots & 0 & 1 \end{bmatrix}. \tag{5.27}$$

The characteristic polynomial of A is $\det(zI - A) = z^n$, where z is a complex variable that should not be confused with the space coordinate. This specific characteristic polynomial has all the eigenvalues placed at the origin of the complex plane. The zeros of the discrete plant transfer function associated to (5.25, 5.26) are given by the roots of the polynomial

$$N(z) = z^{n-1} + \beta z^{n-2} + \ldots + \beta^{n-2} z + \beta^{n-1}. \tag{5.28}$$

These zeros are distributed along a circle of radius β, according to

$$z_i = \beta e^{2\pi i/n}, \quad i = 1, \ldots, n-1. \tag{5.29}$$

It is remarked that the antiresonance pattern associated with the above pole/zero distribution is not due to the discretization procedure but is inherent to the plant structure. The same type of frequency response is also found in the continuous model (Berenguel and Camacho 1996; Meaburn and Hughes 1994).

The controllability (for u and d inputs) and observability matrices are given by

$$\mathcal{C}(A, B) = \begin{bmatrix} 1 & 0 & \ldots & 0 \\ 1 & \beta & 0 & \vdots \\ \vdots & \vdots & \ddots & 0 \\ 1 & \beta & \ldots & \beta^{n-1} \end{bmatrix}, \tag{5.30}$$

$$\mathcal{C}(A, D) = \begin{bmatrix} \beta & 0 & \ldots & 0 \\ 0 & \beta^2 & 0 & \vdots \\ \vdots & 0 & \ddots & 0 \\ 0 & \ldots & 0 & \beta^n \end{bmatrix}, \tag{5.31}$$

$$\mathcal{O}(C, A) = \begin{bmatrix} 0 & \ldots & 0 & 1 \\ \vdots & 0 & \beta & 0 \\ 0 & \ddots & 0 & \vdots \\ \beta^{n-1} & 0 & \ldots & 0 \end{bmatrix}, \tag{5.32}$$

and hence the system is fully controllable and fully observable.

5.3.2 Control Law

The control law is obtained by minimizing in a receding horizon sense the deterministic multistep quadratic cost

$$J_k = \sum_{i=1}^{T} \left[(r(k+i) - y(k+i))^2 + \rho u^2(k+i-1) \right], \tag{5.33}$$

where r is the reference to track that is assumed to be known in advance, and $\rho \geq 0$ is a parameter that weights the penalty on the manipulated variable cost. In order to solve this problem, predictors of y are required. These predictors can be computed using the output Eq. (5.26) to express the prediction of the output in terms of the prediction of the state as

$$y(k+i) = Cx(k+i) \tag{5.34}$$

and then iterating the state Eq. (5.25) to express $x(k+i)$ in $x(k)$ by

$$\begin{aligned} x(k+i) = {} & A^i x(k) + \mathcal{C}_{B_i} \left[u(k+i-1) \ldots u(k)\right]^T \\ & + \mathcal{C}_{D_i} \left[d(k+i-1) \ldots d(k)\right]^T , \end{aligned} \tag{5.35}$$

where, for $i \leq n-1$,

$$A^i = \begin{bmatrix} 0 \ldots 0 & 0 \\ & 0 \\ \beta^i I_{n-1} & \vdots \\ & 0 \end{bmatrix}, \tag{5.36}$$

$$\mathcal{C}_{B_i} = \begin{bmatrix} 1 & 0 & \ldots & 0 \\ 1 & \beta & \ddots & \vdots \\ 1 & \beta & \ddots & 0 \\ \vdots & \vdots & \ddots & \beta^{i-1} \\ 1 & \beta & \ldots & \beta^{i-1} \end{bmatrix} \tag{5.37}$$

and

$$\mathcal{C}_{D_i} = \begin{bmatrix} \beta & 0 & \ldots & 0 \\ 0 & \beta^2 & 0 & 0 \\ \vdots & 0 & \ddots & 0 \\ 0 & \ldots & 0 & \beta^i \\ 0 & \ldots & \ldots & 0 \end{bmatrix}. \tag{5.38}$$

Then, the prediction of future values of the output y comes in a straightforward way by combining the predictions for the state given by (5.35) with the observations model (5.34), to yield

$$y(k+i) = Cx(k+i) = CA^j x(k) + C \sum_{j=0}^{i-1} A^{i-j-1} Bu(k+j). \tag{5.39}$$

The minimization of (5.33) is made under the assumption that both the control action u and the inlet fluid temperature d are constant along the control horizon, meaning that

$$u(k) = \ldots = u(k+i-1) \tag{5.40}$$

and

$$d(k) = \ldots = d(k+i-1). \tag{5.41}$$

In the appendix G.2 it is proved that the control law is then given by

$$u(k) = \frac{\left(\sum_{i=1}^{T} \eta_i\right) r(k) - \sum_{i=1}^{T} \eta_i \beta^i x_{n-i}(k)}{\sum_{i=1}^{T} \eta_i^2 + \rho}, \tag{5.42}$$

where

$$\eta_i = 1 + \beta + \ldots + \beta^{i-1}$$

The following WARTIC-State (WARped TIme controller based on state space model) (Silva et al. 2003a) summarizes the control strategy just described:

WARTIC-State algorithm

At the beginning of each sampling interval, recursively execute the following steps:

1. On the basis of plant measurements made up to the current discrete time k, update the RLS estimates of the parameters α and β in the model (5.13).
2. Compute the virtual manipulated variable $u(k)$ by (5.42).
3. Compute the fluid flow $F(k)$ to apply to the plant from (5.24), by

$$F(k) = \alpha A_f \delta \frac{R_{av}(k)}{u(k)}.$$

4. Compute the duration of the sampling interval during which the flow obtained in the previous step is applied to the plant by

$$\Delta_k = u(k).$$

□

For $\rho = 0$ (no penalty on the control variable) and $T = 1$, the control law (5.42) reduces to

$$u(k) = r - x_{n-1}(k). \tag{5.43}$$

Inserting (5.43) in the state Eq. (5.21) for $j = n$ it is concluded that, with this control law,

$$x_n(k+1) = r.$$

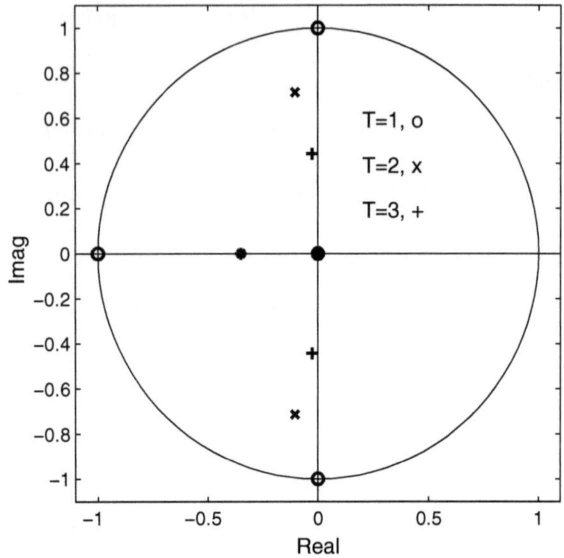

Fig. 5.7 Pole/zero map of the DCSF model controlled with the WARTIC-State algorithm, with the active part of the pipe divided into $n = 4$ segments, showing the migration of the closed-loop poles to the interior of the unit *circle*, and away from its boundary, when the prediction horizon T is increased

i.e. the flow necessary to ensure that the fluid inside the pipe lement volume (element $n - 1$) will reach the output in the next sampling period, with a temperature that is equal to the reference specified.

The choice $\rho = 0, T = 1$ is however a bad choice since it yields closed-loop poles that cancel the plant zeros, causing the occurrence of oscillations in the manipulated variable. By increasing the prediction horizon T these undesirable hidden oscillatory modes disappear. Figure 5.7 provides an example by showing the closed-loop system eigenvalues for $n = 4$ and different values of T. While for $T = 1$ the eigenvalues lie over the stability boundary (the unit circle), and in the same position as the plant zeros, by increasing the predictive horizon T they migrate to the interior of the unit circle, thereby improving relative stability. Since, for $T > 1$, the process zeros are no longer canceled, there are no longer hidden modes.

5.3.3 Observer Design

If the state, given by the fluid temperature along the pipe at points separated by a distance δ, is not directly measured, it can be estimated with an asymptotic observer.

For this sake, an estimate $\hat{x}(k)$ of the state $x(k)$ (that verifies (5.25)), given observations of the output y up to time k, can be recursively updated in time by

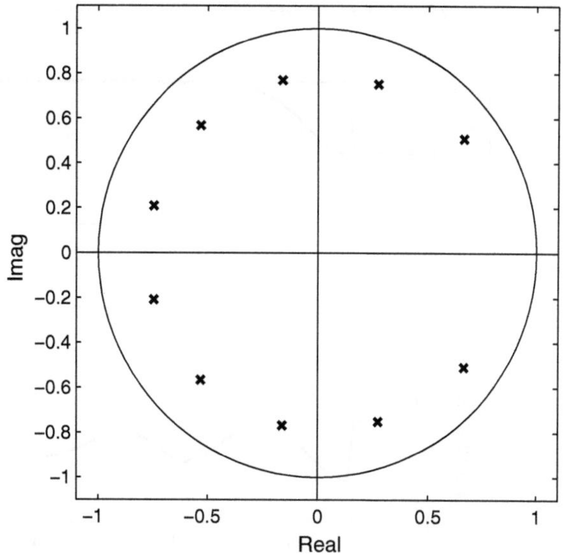

Fig. 5.8 Observer eigenvalues for $n = 10$

$$\hat{x}(k) = A\hat{x}(k-1) + Bu(k-1) + Dd(k-1) \\ + L\left[y(k) - C(A\hat{x}(k-1) + Bu(k-1) + Dd(k-1))\right], \tag{5.44}$$

where L is a vector of observer gains.

The estimation error $\tilde{x}$ defined by

$$\tilde{x}(k) := x(k) - \hat{x}(k), \tag{5.45}$$

satisfies the difference equation

$$\tilde{x}(k) = (A - LC)\tilde{x}(k-1). \tag{5.46}$$

By selecting the vector of observer gains to be

$$L = \left[\tfrac{1}{n} \; \tfrac{2}{n} \; \dots \; \tfrac{n}{n}\right]^T, \tag{5.47}$$

the characteristic polynomial of the observer error Eq. (5.46) becomes

$$P(z) = \det(zI - A + LC) = z^n + \frac{n}{n}z^{n-1} + \frac{n-1}{n}z^{n-2} + \dots + \frac{1}{n}. \tag{5.48}$$

Figure 5.8 shows the distribution of the observer eigenvalues in the complex plane for $n = 10$, as given by the roots of (5.48).

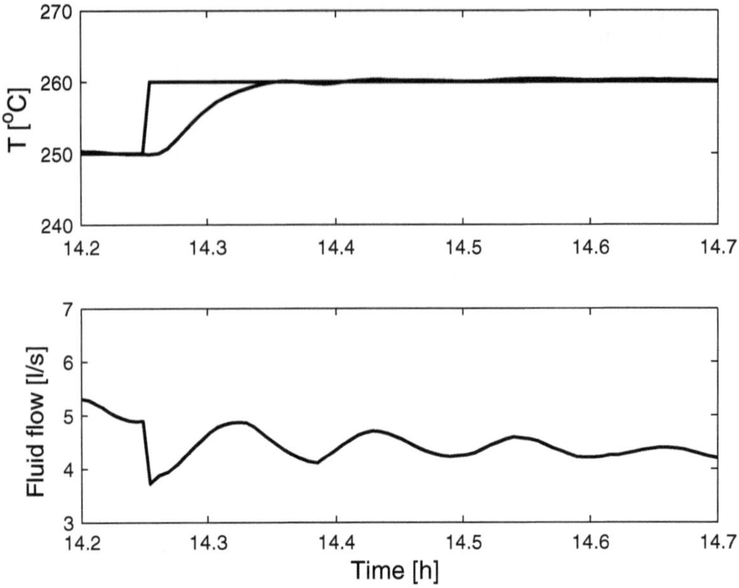

Fig. 5.9 WARTIC-State algorithm. Experiment 1. Response to a temperature step for $T = 8$. Experimental results for the outlet fluid temperature and reference (*above*) and fluid flow (*below*)

5.3.4 Experimental Results with WARTIC-State

Figures 5.9 until 5.11 show experimental results obtained with the WARTIC-*state* algorithm in a DCSF.

An important practical issue is how to select the number n of pipe segments. Together with the range of allowed flows, this parameter establishes the sampling period range. In the DCSF used the total volume V of the active part of the pipe is approximately 1,800 l. The minimum value of the flow is $F_{\min} = 2$ l/s and the maximum is $F_{\max} = 10$ l/s. A value of $n = 20$ was chosen in order to get a sampling period between

$$\Delta^{\min} = \frac{1}{F_{\max}} \frac{V}{n} = 9\,\text{s} \quad \text{and} \quad \Delta^{\max} = \frac{1}{F_{\min}} \frac{V}{n} = 45\,\text{s}$$

In other works performed with the same plant, the values of the sampling period range between 15 and 35 s.

Figure 5.9 documents experiment 1 with WARTIC-State and shows results obtained with $T = 8$. From the plot it is observed that there are strong oscillations in the value of flow that are not reflected in the output, corresponding to an unobservable mode. This undesirable effect is due to the cancelation of antiresonance zeros by the controller poles and, as shown by the subsequent example, can be avoided by increasing the prediction horizon T.

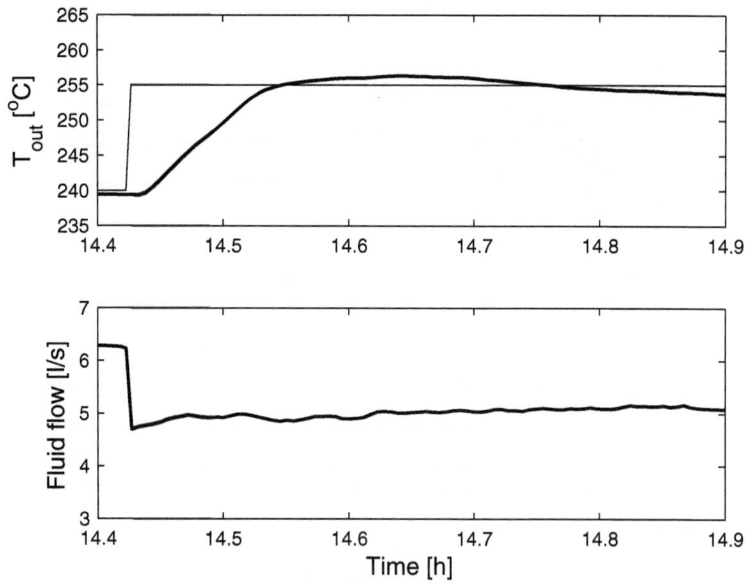

Fig. 5.10 WARTIC-State algorithm. Experiment 2. Outlet fluid temperature and reference (*above*) and fluid flow for $T = 16$. Experimental results

Experiment 2 with WARTIC-State is documented in Figs. 5.10 and 5.11. In this experiment, the prediction horizon has been increased to $T = 16$. Figure 5.10 represents the time evolution of the outlet fluid temperature superimposed on the reference to track and the fluid flow. With this configuration of the controller the fluid temperature is able to change 15 °C in just 6 min. Furthermore, the outflow presents almost no oscillations. The controller is able to perform the required maneuver by taking advantage of the knowledge of the dominant plant dynamics that is embedded in the change in timescale. Furthermore, the use of an extended horizon allows an adequate compensation of the antiresonant peaks in the frequency response, while avoiding the cancelation of the process zeros.

Figure 5.11 shows the estimates of the fluid temperature, in points along the pipe, computed by the state observer.

5.4 Moisture Control

In order to illustrate the nonlinear predictive control design approach in a different plant, consider the moisture drying plant described in Sect. 2.5.1. Figure 5.12 shows simulation results with $T = 3$ and $T = 9$. In both cases the number of segments in the space discretization of the conveyor belt is $n = 10$. The moisture content at the feeding point, $W(0, k)$, is initially equal to 0.5 and it is then made to vary, thereby inducing a disturbance. The desired value for the outlet moisture content is $W(L, k) = 0.1$, where L is the length of the conveyor belt.

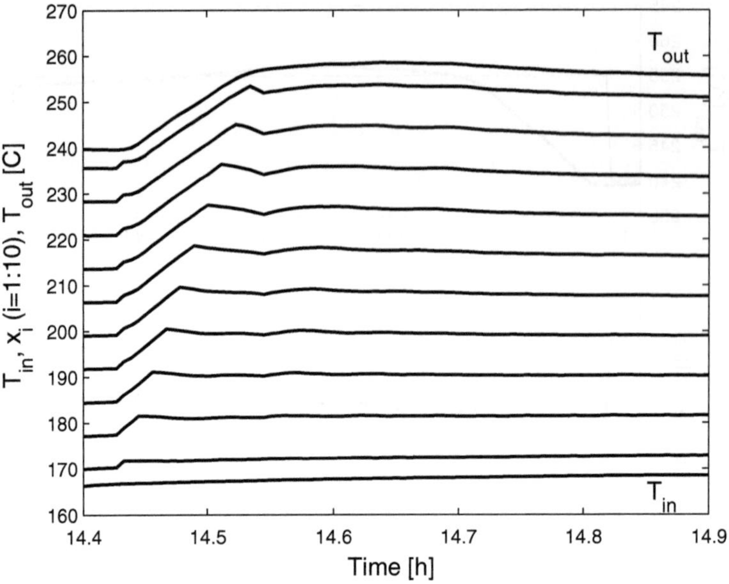

Fig. 5.11 WARTIC-State algorithm. Experiment 2. Estimated temperatures along the pipe for $T = 16$

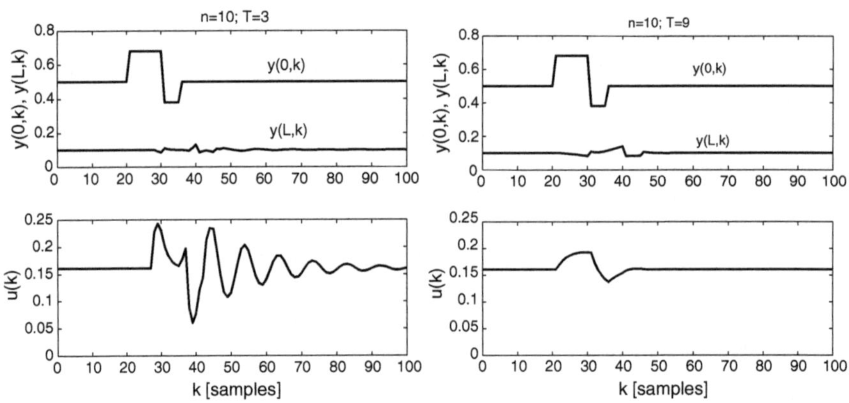

Fig. 5.12 Moisture control with the WARTIC-State algorithm. Simulation results for $T = 3$ (*left*) and $T = 9$ (*right*). In both cases the upper graphics show the moisture at the feeding point (*upper curve*) and the moisture at the belt end (*lower curve*), and the graphic placed below shows the manipulated variable

Comparing the results for $T = 3$ and $T = 9$ it is concluded that the increase in T eliminates the oscillation in the manipulated variable, a situation much similar to the one observed in the control of the DCSF with WARTIC-State presented in the previous section.

5.5 Main Points of the Chapter

The definition of a "warped" timescale in the DCSF model, together with the redefinition of the manipulated variable, yields a transformed model that is exactly linear. Using this linear model, the machinery of linear predictive control can then be used to yield controllers in closed form that minimize a quadratic cost. Two types of controllers are considered, depending on the type of model employed. For input/output models a FIR linear model is obtained. In this case, the resulting controller has the drawback of being a pure feedforward controller. Instead, the state-space formulation yields a feedback/feedforward controller. In both cases it is possible to make sudden jumps of the reference without overshoot. Internal modes, associated to the fluid temperature along the pipe may be highly oscillatory, but the increase of the control horizon prevents this undesirable effect.

In addition to the control of a DCSF, for which experimental data is shown, the same technique is used to control a moisture drying plant.

5.6 Bibliographic Notes

The idea of making a time normalization in order to linearize exactly the model of the DCSF and the use of a variable sampling interval to implement it in practice has been originally introduced in Silva et al. (2003a, b). The construction of the state-space model and the associated state observer is addressed in Silva et al. (2003b). After the change of variable in the timescale, the model becomes linear and the design of the observer is a straightforward modification of the standard procedure (Åstrom and Wittenmark 1997) to include the extra input associated to accessible disturbance. In Pin et al. (2008) an adaptive controller that combines a warped time model with the unscented Kalman filter is applied to a DCSF that employs molten salt as fluid. Although the field reaches higher temperatures than DCSF that use oil as fluid, being therefore close to the thermal limits of the receiver pipes, the algorithm is able to keep the operating point within the region defined by prescribed constraints.

References

Åstrom KJ, Wittenmark B (1997) Computer controlled systems. Prentice Hall, Upper Saddle River

Berenguel E, Camacho EF (1996) Frequency-based adaptive control of systems with antiresonance modes. Control Eng Pract 4(5):677–684

Grüne L, Pannek J (2011) Nonlinear model predictive control. Springer, New York

Johansen TA, Storaa C (2002) Energy-based control of a solar collector field. Automatica 38:1191–1199

Meaburn A, Hughes FM (1994) Prescheduled adaptive control scheme for ressonance cancellation of a distributed solar collector field. Sol Energy 52(2):155–166

Pin G, Falchetta M, Fenu G (2008) Adaptive time-warped control of molten salt distributed collector solar fields. Control Eng Pract 16:813–823

Silva RN, Lemos JM, Rato LM (2003a) Variable sampling adaptive control of a distributed cdollector solar field. IEEE Trans Control Syst Techn 11(5):765–772

Silva RN, Rato LM, Lemos JM (2003b) Time scaling internal predictive control of a solar plant. Control Eng Practice 11(12):1459–1467

Chapter 6
Nonlinear Adaptive Control

A possible approach to the control of nonlinear systems consists of making a change of variables such that they become linear. In Chap. 5 this has been achieved by making a change in the time variable, that is implemented in practice by the use of a time varying sampling interval. In this chapter, instead, the same objective is attained with a change of the manipulated variable, according to a technique known as *feedback linearization*. For that purpose, a virtual manipulated variable is defined, such that, between them and the output dynamics becomes linear.

A linear model may then be used to select the value of the virtual manipulated variable and ensure that the outlet fluid temperature tracks the reference. Inverting the change of variables allows to compute the fluid flow, that forms the physical manipulated variable that is actually applied to the DCSF.

The change of variable that relates the virtual manipulated variable and the fluid flow depends on an unknown parameter that may vary with time. In order to obtain an adaptive controller that tackles this source of uncertainty, a possibility would be to estimate this parameter online from plant data using the RLS estimation algorithm. However, a different approach is followed hereafter that has the advantage to ensure stability of the overall system when the controller is based on parameter estimates (at least assuming the plant nonlinear model to be true). For that sake, a candidate Lyapunov function is defined in the space of the plant states augmented with the states of the estimation algorithm. The estimation algorithm is then designed such that, along the orbits of the augmented system, the total derivative with respect to time of the candidate Lyapunov function is negative semidefinite. This fact ensures stability in the sense of Lyapunov for the overall system (that is to say, to the controlled system, including the estimator). Furthermore, using the invariant set theorem (Slotine 1991), it is possible to conclude that the outlet fluid temperature will tend to the reference when time grows unbound. These conclusions are valid if the nonlinear model assumed for the plant is valid.

Thus, in this chapter, an approach is followed that relies on a combination of feedback linearization and control Lyapunov functions. In the first part of the chapter, a very simple (first order) nonlinear model is considered. Upon linearization using the technique mentioned, a single integrator results that can be controlled with a constant

J. M. Lemos et al., *Adaptive Control of Solar Energy Collector Systems*, Advances in Industrial Control, DOI: 10.1007/978-3-319-06853-4_6,

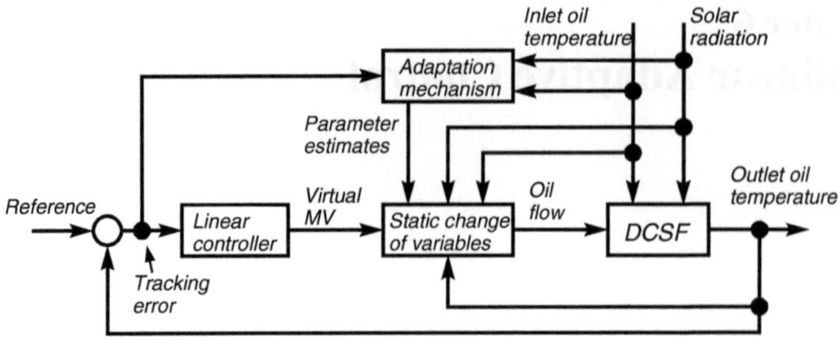

Fig. 6.1 Structure of the nonlinear adaptive control with a reduced complexity model

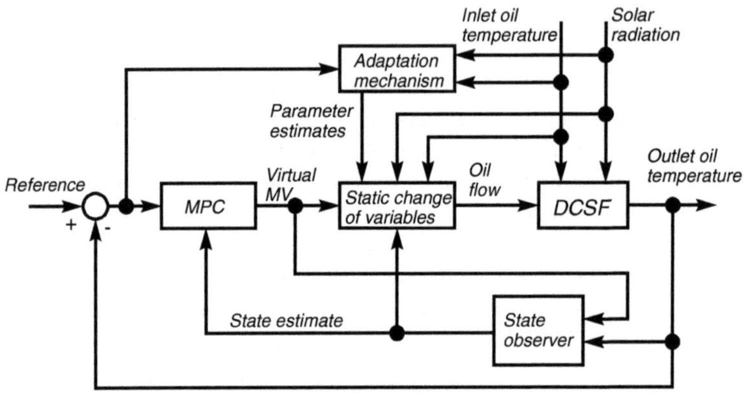

Fig. 6.2 Structure of the nonlinear adaptive MPC control with a higher order model

gain. The simplicity of the model has the advantage of greatly simplifying the design to better illustrate the technique. Despite this simplification, the conclusions are valid on an actual DCSF, and experimental results are presented to illustrate this fact. The block diagram is shown in Fig. 6.1.

In the second part of the chapter, a more general situation is considered, corresponding to the block diagram of Fig. 6.2. The dynamics of the DCSF is no longer represented by a first-order model but, instead, by a n-th order lumped parameter model whose state is made of the fluid temperature at points along the pipe. Furthermore, using a bounding technique described in Primbs (2000), Sontag (1998), it is possible to define a receding horizon controller (RHC) that ensures stability of the nonlinear lumped parameter DCSF model.

6.1 Feedback Linearization

Feedback linearization is a technique that consists of making a change of variables such that, in the new coordinates, the system becomes exactly stable (Slotine 1991; Isidori 1995). Depending on the variables involved, there are two types of feedback linearization procedures, namely input to state linearization and input–output linearization. It is possible to show that the lumped parameter DCSF model (2.19) is input to state linearizable only for $n = 1$ and $n = 2$. Hence, input to state linearization is not further pursued here. Instead, input–output exact linearization is considered hereafter.

6.1.1 Input–Output Linearization

Input–output exact linearization consists of finding a state transformation (that is a diffeomorphism, i.e., an invertible function that maps onde differentiable manifold to another one, and the function and its inverse are invertible)

$$z = \Phi(x) \tag{6.1}$$

and an input transformation

$$u = \frac{v - b(x)}{a(x)}, \tag{6.2}$$

where v is the transformed input and $a(.)$, $b(.)$ are convenient state functions (to be made precise below), such that the transformed system has the form

$$\begin{aligned} \dot{z}_1 &= z_2 \\ \dot{z}_2 &= z_3 \\ &\vdots \\ \dot{z}_{n_r-1} &= z_{n_r} \\ \dot{z}_{n_r} &= v \\ \dot{z}_{n_r+1} &= q_{n_r+1}(z) \\ &\vdots \\ \dot{z}_n &= q_n(z) \\ y &= z_1, \end{aligned} \tag{6.3}$$

where $y = h(x)$ is the system output and n_r is the so called *relative degree*. The relative degree (Slotine 1991; Isidori 1995) is given by the number of times the output y has to be differentiated at a given time t_0 so that $u(t_0)$ explicitly appears.

In this transformed form, the relation between the new input v and the output is an integrator of order n_r. The last $n - n_r$ equations correspond to unobservable states and they must be shown to be stable.

Consider now the DCSF model defined by: (2.19) where, for the sake of simplicity and without loss of generality, it is assumed that the inlet fluid temperature is

$T(0, t) = 0$. Define the output y by:

$$y = h(x) = x_n. \tag{6.4}$$

Differentiating the output given by (6.4) one obtains

$$\dot{y} = \dot{x}_n = \alpha R - \frac{x_n - x_{n-1}}{h} u \tag{6.5}$$

and it is concluded from the definition that the relative degree of this system is $n_r = 1$. The transformation $z = \Phi(x)$ that brings the system to the form (6.3) is thus of the form

$$z = \Phi(x) = \begin{bmatrix} h(x) \\ \phi_2(x) \\ \vdots \\ \phi_n(x) \end{bmatrix}, \tag{6.6}$$

with convenient functions $\phi_2, \ldots, \phi_n$.

Consider the control law given by:

$$u = \frac{-L_f h(x) + v}{L_g h(x)} = \frac{\alpha R - v}{x_n - x_{n-1}} h, \tag{6.7}$$

where v is a virtual manipulated variable. In (6.7), the symbol $L_g h(x)$ is the Lie derivative of h defined by

$$L_g h(x) \triangleq \frac{\partial h}{\partial x} g(x). \tag{6.8}$$

Equation (6.7) provides a transformation such that, from the transformed input v (the virtual manipulated variable) to the output y, the dynamics reduces to a pure integrator. Once the value of v is computed with an appropriate control law, the actual physical manipulated variable u to apply to the plant is computed from v by (6.7).

It is remarked that the computation of u requires, in addition of v, the values of the states x_n and x_{n-1} that must be available for measurement. The measure of x_{n-1} is not available in the plant considered for values of $n > 1$. In general, one possibility is to estimate this variable. Another line of work, exploited hereafter, consists in using approximate models.

6.1.2 Internal Dynamics

With the above input–output feedback linearization control strategy, it remains to show that the internal dynamics (i.e., the states that are not observable from the

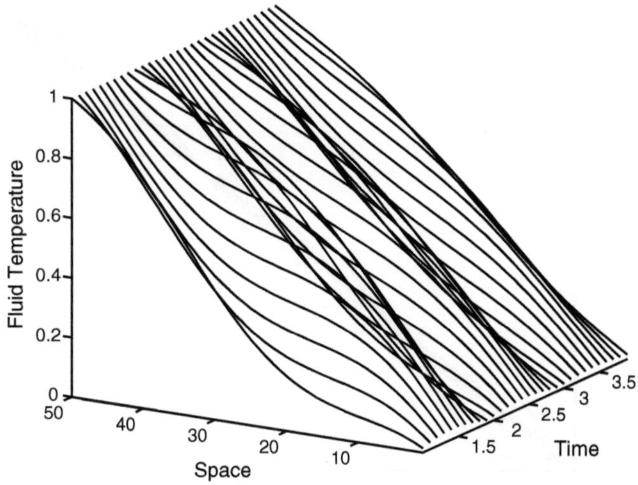

Fig. 6.3 Example of internal dynamics of a DCSF model with input–output feedback linearization control. Normalized variables are used

output) is stable. For a DCSF, a convenient way for studying internal dynamics is by considering tracking dynamics (Isidori 1995). Let the initial condition $x^0 = x(0)$ be compatible with the reference signal $y_r(t)$. This means that

$$\begin{aligned} y_r(0) &= h(x^0) \\ y_r^{(1)}(0) &= L_f h(x^0) \\ &\vdots \\ y_r^{(r-1)}(0) &= L_f^{r-1} h(x^0). \end{aligned}$$

Define the *tracking manifold* by:

$$M_t = \left\{ x \in M \; : \; h(x) = y_r(t), \; \ldots, \; L_f^{r-1} h(x) = y_r^{(r-1)}(t) \right\},$$

where M denotes the state-space.

The tracking dynamics is the dynamics of the system constrained to M_t. It corresponds to the internal dynamics when the output is perfectly tracking the reference $y^*(t)$. In general, finding the internal dynamics involves computing Φ that may require a cumbersome numerical computation.

Figure 6.3 shows the internal dynamics when the initial state is a straight line added to a sinusoid with a period equal to the length of the pipe. Normalized variables are used. As seen in this picture, although the input and output temperatures are constant, intermediate values of the state present a noticeable oscillation. By choosing another initial condition, a different internal dynamics is yielded. Figure 6.4 shows another example in that the oscillations have a spatial period equal to $L/3$.

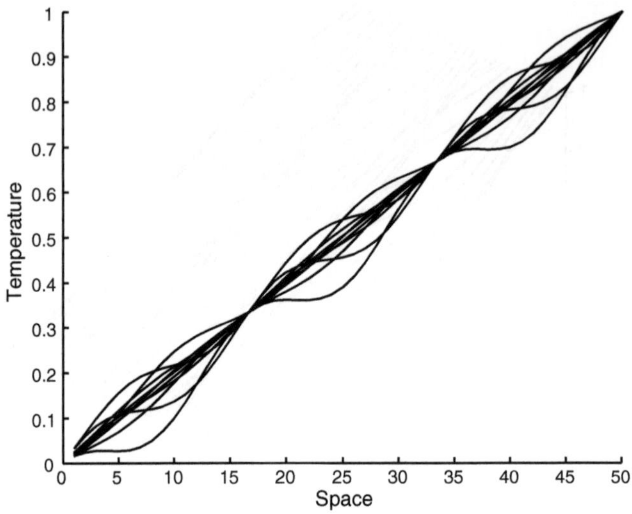

Fig. 6.4 Another example of internal dynamics of a DCSF model with input–output feedback linearization control. Each *curve* corresponds to the spatial temperature distribution at a different time. Normalized variables are used

The following result that characterizes the internal dynamics of the DCSF lumped parameter model is proved in the Appendix G.

Proposition 6.1 *Let the length L of the active part of the DCSF be normalized such that $L = 1$ and assume there are no losses ($\gamma = 1$ in (2.15)). The eigenvalues of the linearized tracking dynamics of the field when the state is constrained by:*

$$\dot{x}_n = 0, \quad x_n = r = 1, \tag{6.9}$$

where $y^* = 1$ is the reference to track and n is the number of segments in the space grid, are placed equally spaced over a circumference of radius n and centered at $-n$. When n grows, these eigenvalues approach the imaginary axis. □

6.1.3 Internal Dynamics with Approximate Control Laws

In a DCSF, the change of input variable defined by (6.7), according to which the physical manipulated variable (fluid flow) u is replaced by the virtual manipulated variable v, leads to a relation between the transformed input v and the output y exactly given by an integrator. It happens, however, that in a DCSF, while $y = x_n$ is available for measurement, x_{n-1} is in general not. Indeed, only the inlet and outlet fluid temperatures are available for measurement.

In this realm, one issue with practical incidence to consider is plant internal dynamics when (6.7) is replaced by:

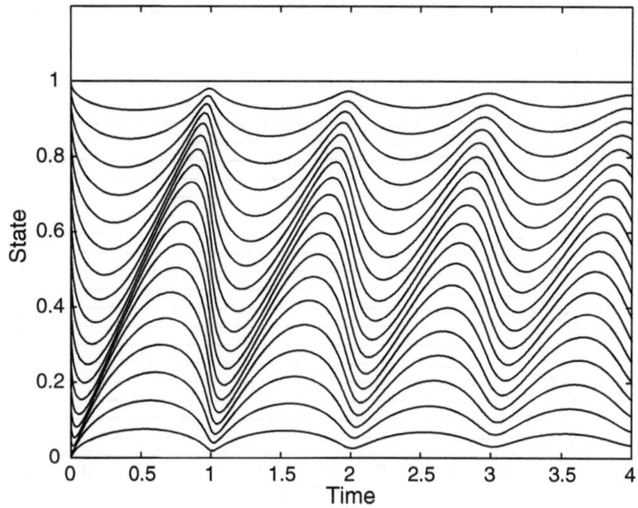

Fig. 6.5 Internal dynamics with $d = L/n$, corresponding to exact feedback linearization. Normalized variables are used

$$u = \frac{\alpha R - v}{T(L, t) - T(L - d, t)}, \tag{6.10}$$

where d is the distance between both temperature sensors that are assumed to exist.

Figures 6.5 and 6.6 correspond to the situations in which $d = L/n$ (corresponding to exact input–output linearization) and $d = L$ (corresponding to an approximation of the feedback linearization law). Each of these figures shows the time evolution of the state components, starting from a given initial condition.

Start by considering Fig. 6.5. The output temperature, corresponding to the component initialized at 1 (i. e., the upper curve) is constant. This is what is expected, since $v = 0$ and the relation between v and y is exactly an integrator. The other states, according to the previous discussion on internal dynamics, present oscillations.

Consider now Fig. 6.6. The output is no longer constant and there is an initial transient which is explainable by the fact that the control transformation used is now only an approximation to the linearizing one.

6.2 Control with a Reduced Complexity Model

Motivated by the discussion on the previous section, control is now designed on the basis of the reduced complexity model

$$\dot{y} = -u(y - y_0)\frac{1}{L} + \alpha R, \tag{6.11}$$

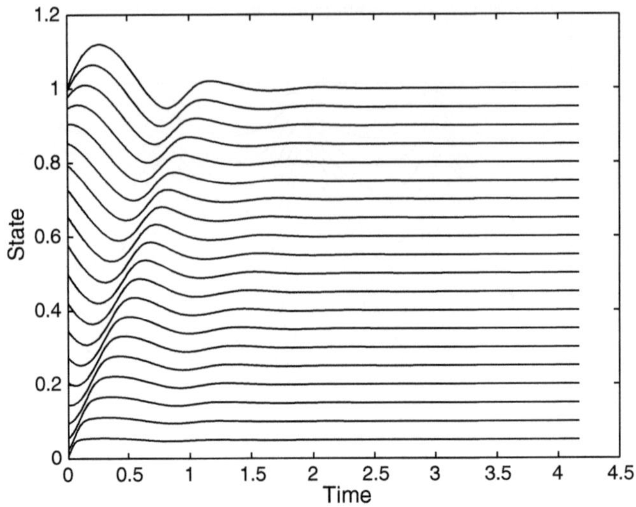

Fig. 6.6 Internal dynamics with $d = L$, corresponding to an approximation of the feedback linearization law. Normalized variables are used

where y is the outlet fluid temperature, y_0 is the inlet fluid temperature and u is the fluid flow velocity. This model corresponds to the lumped parameter approximation 2.19 with $n = 1$.

6.2.1 Feedback Linearization of the Reduced Complexity Model

In (6.11), the optical efficiency parameter α is not exactly known in advance and, furthermore, it includes modeling errors, being expected to change slowly by factors, discussed in broad terms in Sect. 1.3, such as dust accumulation or mirror deformations due to wind. It is thus to be estimated. Let α denote the true optical efficiency, $\hat{\alpha}$ an estimate of α and $\tilde{\alpha}$ the error between the two

$$\tilde{\alpha} = \alpha - \hat{\alpha}. \tag{6.12}$$

Then (6.11) yields

$$\dot{y} = -u(y - y_0)\frac{1}{L} + (\hat{\alpha} + \tilde{\alpha})R. \tag{6.13}$$

From (6.7) with $d = L$, the virtual control signal v is given by

$$v = -u(y - y_0)\frac{1}{L} + \hat{\alpha}R. \tag{6.14}$$

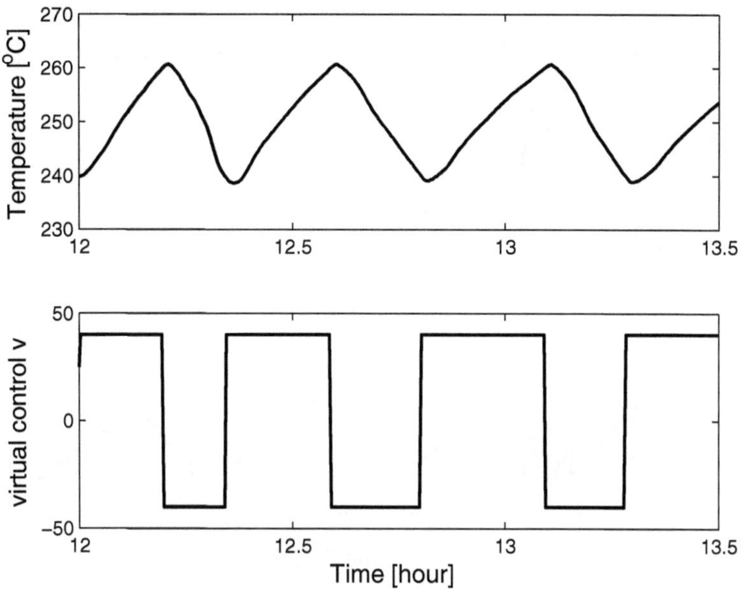

Fig. 6.7 Experimental results on input–output linearizing control of a DCSF using a reduced complexity model. Outlet fluid temperature (*above*) and virtual control signal (*below*)

This expression may now be used to control the outlet fluid temperature by using a linear controller. The actual control signal applied to the plant is given by:

$$u = \frac{\hat{\alpha}R - v}{y - y_0}L. \tag{6.15}$$

Inserting (6.15) in (6.13) yields

$$\dot{y} = v + \tilde{\alpha}R. \tag{6.16}$$

Equation (6.15) defines the static transformation Ψ such that between the variables v and y the model dynamics is equivalent to a linear system (integrator), disturbed by an input offset with an unknown gain.

Figures 6.7 and 6.8 show experimental results in which a DCSF has been coupled with the approximate feedback linearizing law (6.15) based on the reduced complexity model (6.11). The objective is to illustrate that, in a DCSF, from the virtual control signal v to y, the system behaves approximately as an integrator, even in the presence of strong disturbances in solar radiation. In this experiment, only the linearizing loop is closed. The system (DCSF plant plus the feedback linearizing loop) is driven by a virtual manipulated variable signal formed by a sequence of steps with varying amplitudes and duration depicted in Fig. 6.7. As expected, a saw-tooth like temperature signal results by integrating the piecewise constant virtual input.

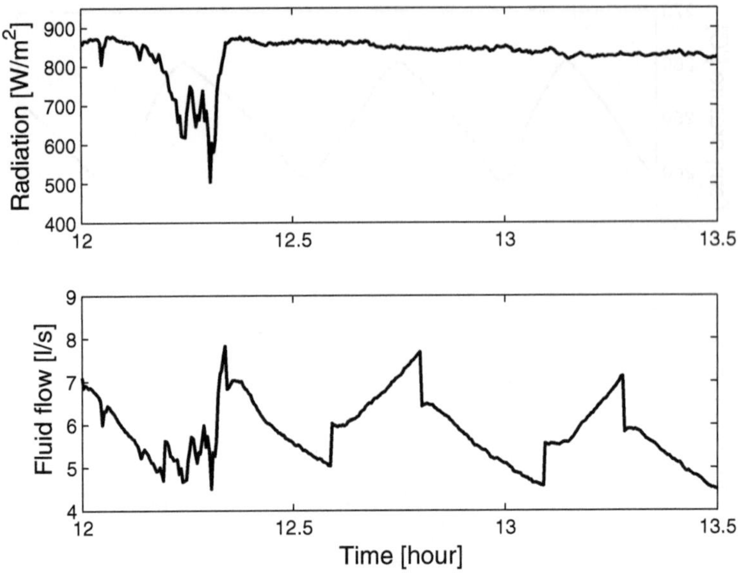

Fig. 6.8 Experimental results on input–output linearizing control of a DCSF using a reduced complexity model. Solar radiation (*above*) and the fluid flow (*below*)

This behavior is actually observed, as seen in Fig. 6.7 (above) that shows the outlet fluid temperature varying approximately as a triangular signal with different slopes depending on the amplitude of the virtual control (Fig. 6.7, below).

Some discrepancies are explainable by the error in the value of the estimate used for α, in accordance with Eq. (6.16). By using a better estimate, as would happen if an adaptive scheme is used, an even better approximation to the integrator would be obtained.

It can also be seen that the sudden reduction in solar radiation, represented in Fig. 6.8 (above)) is almost not sensed in the output since the actual control signal, shown in Fig. 6.8 (below), compensates its wild changes.

6.2.2 Control Lyapunov Function-Based Adaptation

In order to obtain an adaptation law for updating $\hat{\alpha}$ and incorporating this estimate in an adaptive controller that ensures stability of the nominal model, an argument based on the Second Lyapunov's Method (Slotine 1991) is used. For that sake, consider the candidate Lyapunov function defined by:

$$V(e, \tilde{\alpha}) = \frac{1}{2}\left(e^2 + \frac{1}{\gamma}\tilde{\alpha}^2\right), \tag{6.17}$$

where $\gamma > 0$ is a constant parameter and e is the tracking error defined by:

$$e(t) = y^* - y(t), \tag{6.18}$$

with y^* being a constant set-point. Since the linearized system is an integrator, there is no need to include integral action in the controller. Thus, let the virtual control in continuous time be given by the Proportional Derivative (PD) law

$$v = k_p e - k_d \dot{y}, \tag{6.19}$$

with k_p and k_d constant gains. Replacing this control law in the transformed plant Eq. (6.16), yields the following closed-loop model

$$\dot{y} = \frac{k_p}{1+k_d} e + \frac{R}{1+k_d} \tilde{\alpha}. \tag{6.20}$$

For V to be a Lyapunov function, its time derivative must be semidefinite negative. Assuming a constant set-point and a constant α, use the parameter estimation error definition (6.12), and get

$$\dot{V} = -e\dot{y} - \frac{1}{\gamma} \tilde{\alpha} \dot{\hat{\alpha}}. \tag{6.21}$$

Upon using (6.20), this becomes

$$\dot{V} = -\frac{k_p}{1+k_d} e^2 - \left[\frac{R}{1+k_d} e + \frac{1}{\gamma} \dot{\hat{\alpha}} \right] \tilde{\alpha} \tag{6.22}$$

The adaptation law for $\hat{\alpha}$ is chosen such that the term multiplying $\tilde{\alpha}$ in the above expression vanishes, yielding

$$\dot{\hat{\alpha}} = -\frac{\gamma}{1+k_d} Re. \tag{6.23}$$

With this choice, $\dot{V}$ is negative semidefinite if $k_p > 0$ and $k_d > -1$. Furthermore, by LaSalle's invariance theorem (Slotine 1991), all the trajectories converge to the maximum invariant set contained in the set of states where $\dot{V} = 0$, implying that

$$\lim_{t \to \infty} y(t) = y^*, \tag{6.24}$$

that is to say, with the feedback linearization rule (6.15), where $\hat{\alpha}$ is updated with the adaptation law (6.23), and the feedback control that computes the manipulated variable v according to (6.19), the outlet fluid temperature will tend to the reference (assumed to be constant).

It is remarked that this conclusion applies only to the simplified model (6.11) that is only an approximation to the actual DCSF dynamics. However, experimental results show that the resulting performance is acceptable, and in accordance with the assumptions made. Indeed, experimental results on the input–output linearizing control using a constant estimate of the mirror efficiency α have already been shown in the previous section. Results with the full adaptive controller are presented hereafter.

Observe also that nothing is said about the convergence of the estimate $\hat{\alpha}$ to its true value α. By extending the argument on LaSalle's invariance theorem it is possible to show that, if the reduced complexity model is used to represent the DCSF dynamics, $\hat{\alpha}$ will converge to α provided that the temperature reference does not vanish.

6.2.3 Experimental Results with Reduced Complexity Control

The nonlinear adaptive control law described in Sect. 6.2.2 is illustrated hereafter in an experiment in which a feedback linearizing loop based on the reduced complexity model (6.11) is used and then an outer-loop is closed so as to track a constant temperature reference while the mirror efficiency parameter α is estimated using the approach just explained based on a joint Lyapunov function for control and estimation.

The virtual control signal v is generated by the following discrete time finite difference approximation of the PD controller (6.19)

$$v(k) = k_p e(k) + k_d(y(k) - y(k-1)). \tag{6.25}$$

The manipulated variable signal that is actually applied to the plant (fluid flow) is computed from v by:

$$u(k) = \frac{\hat{\alpha}(k)R(k) - v(k)}{y(k) - y_0(k)} h, \tag{6.26}$$

with the estimate $\hat{\alpha}$ updated by the following approximation to (6.23)

$$\hat{\alpha}(k+1) = \hat{\alpha}(k) - k_a Re(k). \tag{6.27}$$

A sampling time of 15 s and the following gains are used:

$$k_p = 9, \quad k_a = 0.0001/820, \quad k_d = 15.$$

During the experiment the radiation was roughly constant, its mean value of 820 Wm^{-2} being lumped in the constant k_a.

Figure 6.9 (above) shows the reference temperature and average outlet temperature taken over all the ten field loops. The overshoot in positive steps can be reduced by using lower gains on the PD controller. Figure 6.9 (below) shows the actual control (fluid flow) applied to the plant.

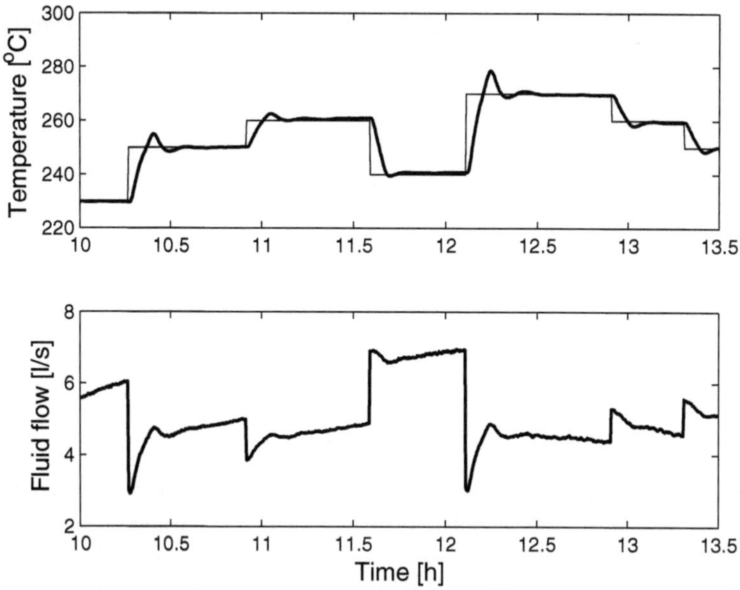

Fig. 6.9 Experimental results on input-output linearizing control of a DCSF using a reduced complexity model. Reference temperature and average outlet temperature (*above*) and the fluid flow applied to the plant (*below*)

6.3 Higher Order Models and Receding Horizon Control

We now generalize the methods of Sect. 6.2 to higher order models. Again, the adaptive control algorithm comprises three main blocks: A receding horizon controller, a control Lyapunov function-based adaptation rule, and a state observer. Furthermore, in order to obtain a receding horizon controller that is guaranteed stable, an extra condition is included as in Primbs (2000). Each of these are detailed hereafter.

6.3.1 *Nonlinear Adaptive Control with Higher Order Models*

Define again the feedforward control law given by:

$$u^* = \frac{\alpha \; R^* \; L}{y^* - x_0}, \tag{6.28}$$

where y^* is the set-point of the outlet fluid temperature x_n^* and R^* is the solar radiation for which the equilibrium is computed. It is assumed that the set-point y^* is always selected such that $y^* > x_0$, and implying that $u^* > 0$. Let x^* be the corresponding equilibrium state and consider the dynamics of the error

$$e = x - x^*. \tag{6.29}$$

The dynamics if e is obtained by subtracting Eq. (2.19) and

$$\dot{x}^* = -\frac{u^*}{L}\left(Ax^* + Bx_0\right) + C\ \alpha\ R^*. \tag{6.30}$$

The error dynamics is given by:

$$\dot{e} = \frac{-A}{L}\ u^*\ e + \frac{-A\ e - A\ x^* - B\ x_0}{L}\ \tilde{u} + C\ \alpha\ \tilde{R}, \tag{6.31}$$

where $\tilde{u} = u - u^*$ and $\tilde{R} = R - R^*$ are deviations with respect to the equilibrium values. As shown in Appendix H, for $\tilde{u} = 0$ and $\tilde{R} = 0$ the error dynamics is asymptotically stable whenever the matrix $\bar{A} = -\frac{A}{L}$ is asymptotically stable or.

Since $-A$ is asymptotically stable, there are matrices $P = P^T \succ 0$, $M_0 = M_0^T \succ 0$ that satisfy the Lyapunov equation

$$(-A)^T P + P(-A) = -M_0. \tag{6.32}$$

Assume the external signal R to be such that $\tilde{R} = 0$. In order to obtain a stabilizing control law for (6.31), consider the following scalar function with P verifying (6.32)

$$V_0 = e^T Pe. \tag{6.33}$$

Its derivative is given by:

$$\dot{V}_0 = -\frac{\alpha R(t)}{r_* - x_0} e^T(A^T P + PA)e + \tilde{u}^T G^T(e, x_0)Pe + e^T PG(e, x_0)\tilde{u}, \tag{6.34}$$

with

$$G(e, x_0) := -\frac{A\ e + A\ x_* + B\ x_0}{L}.$$

Let the control be given by:

$$u(t) = u^*(t) + \tilde{u}(t), \tag{6.35}$$

with

$$\tilde{u} = -\frac{1}{2}\ \phi\ G^T(e, x_0)Pe, \tag{6.36}$$

where $\phi > 0$ is a parameter knob. Under this control law, (6.34) becomes

$$\dot{V}_0 = -e^T \left(M + \phi P G(e, x_0) G^T(e, x_0) P \right) e < 0 \quad \forall e \neq 0, \tag{6.37}$$

where

$$M(t) = \frac{\alpha R(t)}{y_* - x_0} M_0. \tag{6.38}$$

Therefore, V_0 is a Lyapunov function for the error dynamics (6.31) when the loop is closed with (6.35) with $\tilde{u}$ given by (6.36), thereby establishing asymptotic stability when the control law (6.36) is used.

6.3.2 Stabilizing Receding Horizon Control

The above stabilizing controller is now used to build a constraint that ensures stability in a receding horizon controller (RHC). For that sake, define the RHC for the error dynamics by solving the minimization problem

$$\min_u J = \int_t^{t+T} \left(e^T(\tau) P e(\tau) + \gamma\, \tilde{u}^2(\tau) \right) d\tau \tag{6.39}$$

subject to

$$\dot{e} = \frac{-A}{L}\, u^*\, e + \frac{-A\, e - A\, x^* - B\, x_0}{L}\, \tilde{u} \tag{6.40}$$

and

$$V_0(t+T) \geq V_{rhc}(t+T), \tag{6.41}$$

in which y^* is given by (6.28),

$$\begin{aligned} V_0(T) &= e^T(T) \mid_{\tilde{u}=0} P e(T) \mid_{\tilde{u}=0} \\ &= e(0)^T \left(e^{\frac{-A}{L} \int_t^{t+T} u^*(\tau) d\tau} \right)^T P \left(e^{\frac{-A}{L} \int_t^{t+T} u_*(\tau) d\tau} \right) e(0), \end{aligned} \tag{6.42}$$

and

$$V_{rhc}(T) = e^T(T) P e(T), \tag{6.43}$$

where P is computed from (6.32).

The constraint (6.41) is equivalent to impose to the RHC that, at each iteration, the norm of the error at the end of the optimal sequence is bounded by the same norm resulting from the error when u is given by the stabilizing control law (6.35). The existence of a control law stabilizing the closed loop allows to interpret V_0 as a Control Lyapunov Function (Sontag 1998) and is a sufficient condition to ensure Global Asymptotic Stability of the loop closed by the RHC (Primbs 2000).

6.3.3 State Observer

When the state is not available for direct measurement, it must be estimated. For that purpose, associate to (2.19) the following state estimator with output error reinjection, where $\hat{x}$ is the estimate of x that satisfies

$$\dot{\hat{x}} = -\frac{u}{L}\left(A\hat{x} + Bx_0\right) + C\hat{\alpha}R(t) + K(t)(y - D\hat{x}), \tag{6.44}$$

with the measurement model

$$y = D\hat{x} = \begin{bmatrix} 0 & 0 & \cdots & 1 \end{bmatrix} x. \tag{6.45}$$

The observer error dynamics $e_1 := x - \hat{x}$ verifies

$$\dot{e_1} = -\frac{u}{L}\, A\, e_1 + C\, (\alpha - \hat{\alpha})\, R(t) - K(t)D\, e_1, \tag{6.46}$$

with the observer gain $K(t)$ suitably defined as discussed below.

6.3.4 Control Lyapunov Function-Based Adaptation

Consider the candidate Lyapunov function

$$V_1 = e_1^T Q e_1 + \frac{1}{\gamma}\, \tilde{\alpha}^2, \tag{6.47}$$

where $\gamma > 0$ is a parameter, Q is a positive definite matrix and the parameter estimation error $\tilde{\alpha}$ is defined as:

$$\tilde{\alpha}(t) := \alpha - \hat{\alpha}(t), \tag{6.48}$$

where $\hat{\alpha}$ is the estimate of α.

Defining

$$A_e := -\frac{u}{L}A - K(t)D, \tag{6.49}$$

Equation (6.46) reads

$$\dot{e_1}(t) = A_e e_1 + C\tilde{\alpha}R(t). \tag{6.50}$$

The derivative of the candidate Lyapunov function V_1 is then given by:

$$\dot{V}_1 = e_1^T(A_e^T Q + QA_e)e_1 + 2\tilde{\alpha}(CR(t))^T Qe_1 + \frac{2}{\gamma}\tilde{\alpha}\,\dot{\tilde{\alpha}}. \tag{6.51}$$

Stability holds if

$$-M(t) = \left(A_e^T Q + QA_e\right) < 0$$

and

$$\dot{\tilde{\alpha}} = -\rho(CR(t))^T Qe_1$$

from which the following adaptation law follows

$$\dot{\hat{\alpha}} = \rho(CR(t))^T Qe_1. \tag{6.52}$$

It is remarked that, if the pair $(A,\ D)$ is observable, the condition $M(t) \succ 0$ is ensured by the following choice of the observer gain

$$K(t) = \frac{u}{L}K_0, \tag{6.53}$$

where K_0 is selected such that $-A - K_0D$ is asymptotically stable. Indeed, since the fluid velocity u is bounded above and below by strictly positive values

$$u_{\max} \geq u \geq u_{\min} > 0, \tag{6.54}$$

this choice of $K(t)$ yields

$$\begin{aligned} A_e^T Q + QA_e &= \frac{|u|}{L}\left\{(-A - K_0D)^T Q + Q(-A - K_0D)\right\} \\ &= -\frac{|u|}{L}M_0 = -M(t) \end{aligned} \tag{6.55}$$

with the matrix M_0 selected as the positive definite solution of the Lyapunov equation

$$(-A - K_0D)^T Q + Q(-A - K_0D) = -M_0. \tag{6.56}$$

This ensures stability because, from (6.55) and the adaptation law (6.52):

$$\dot{V}_1 = -e_1^T M(t)e_1 = -\frac{|u|}{L}e_1^T Me_1 \leq -\frac{u_{\max}}{L}\,e_1^T Me_1 \leq 0. \tag{6.57}$$

From La Salle's Invariance Principle (Slotine 1991), it is concluded that all state variables will converge to the greatest invariant set contained in the set in which $\dot{V}_1 = 0$. This set is not empty and, furthermore, from (6.57), it is concluded that all the points in it are characterized by the fact that $e_1 = 0$. Hence it follows that $\lim_{t\to 0} e_1(t) = 0$.

6.3.5 Computational Algorithm

A computational efficient version of the the adaptive RHC is given by:

$$\min_{u_1,\ldots,u_{N_u}} J = \int_t^{t+T} \left(e^T(\tau)Pe(\tau) + \gamma\tilde{u}^2(\tau)\right) d\tau \tag{6.58}$$

subject to

$$\begin{aligned} \dot{x} &= -\frac{u}{L}\ (Ax + Bx_0) + C\ \hat{\alpha}\ R(t) \\ x(t) &= \hat{x}(t) \\ u_{\text{seq}}(\bar{t}) &= \text{seq}\{u_1, \ldots, u_{N_u}\} \\ u_{\max} \geq u &\geq u_{\min} > 0 \\ V_0(x(t+T)) &\geq V_{\text{rhc}}(xt+T)) \end{aligned}$$

with V_0 computed by (6.42) with x and α replaced by their estimates.

The estimate of u^* is given by:

$$\hat{u}^* = \frac{\hat{\alpha}\ R(t)\ L}{r(t) - x_0(t)}. \tag{6.59}$$

Here, $\hat{x}$ and $\hat{\alpha}$ are obtained using the state estimator (6.44), the adaptation law (6.52), $u_{\text{seq}}(\bar{t})$ is a sequence of step functions with amplitude u_i $(i = 1, \ldots, N_u)$ and duration $\frac{T}{N_u}$. The variable $\bar{t}$ represents time during the minimization horizon $\bar{t} \in [0, T]$.

Once the minimization result $u(\bar{t})$ is obtained, according to a receding horizon scheme u_1 is applied to the plant at $t+\delta$ and the whole process is restarted, δ being an interval of time which is at least the time needed to compute the solution (Findeisen and Allgöwer 2002).

If the minimization is not feasible, than one can still apply to the plant the control given by:

$$u(t) = u^*, \tag{6.60}$$

that preserves the closed-loop stability.

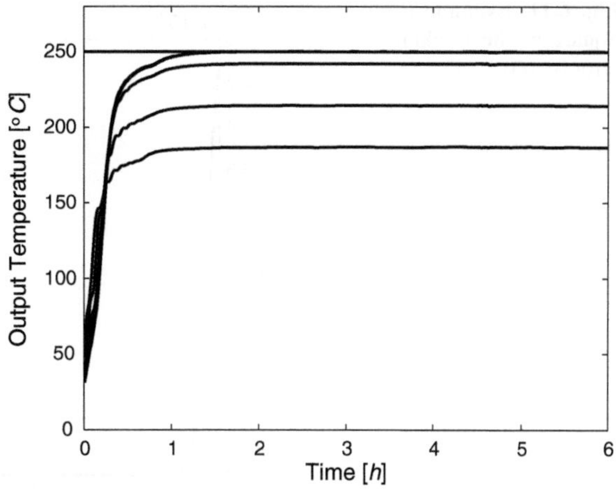

Fig. 6.10 Example 6.1. Reference ($y^* = 250\,°\text{C}$) and state temperatures

6.3.6 Control of a DCSF

In order to illustrate the above receding horizon controller, simulation results of the RHC described in Sect. 6.3.5 are shown hereafter in a detailed nonlinear model of the solar field using experimental sequences for $R(t)$ and $T_0(t)$.

The reduced model uses orthogonal collocation with the 3 interior collocation points $z = [0.113 \;\; 0.500 \;\; 0.887]$. Matrix A is given by

$$A = \begin{bmatrix} 3.87298 & 2.06559 & -1.29099 & 0.67621 \\ -3.22749 & 0.00000 & 3.22749 & -1.50000 \\ 1.29099 & -2.06559 & -3.87298 & 5.32379 \\ -1.87836 & 2.66667 & -14.78831 & 13.00000 \end{bmatrix}$$

and is easy to check that the eigenvalues of $-A/L$ have all negative real part.

In order to configure the controller, the following parameter choices have been made:

$$\gamma = 1 \times 10^5$$
$$K_0 = [15\; 15\; 15\; 15]^T$$
$$\rho = 1.5 \times 10^{-10}$$

Example 6.1 The first example illustrates the fact that the control law (6.35) actually yields a stable closed loop. Figures 6.10 through 6.11 depict the corresponding closed-loop behavior when $r^* = 250\,°\text{C}$. Figure 6.10 shows the reference, constant and equal to 250° and the temperature at the collocation points that form the state of the finite-dimensional approximation used. As expected, the outlet temperature approaches the reference, while the other temperature converge to constant values

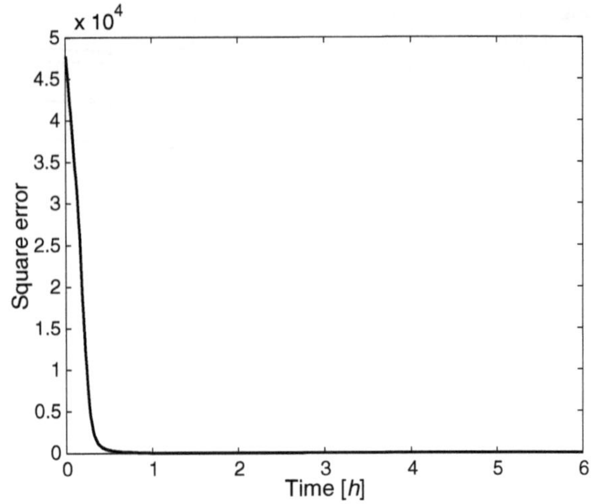

Fig. 6.11 Example 6.1. Square of the tracking error ($e^2(t)$)

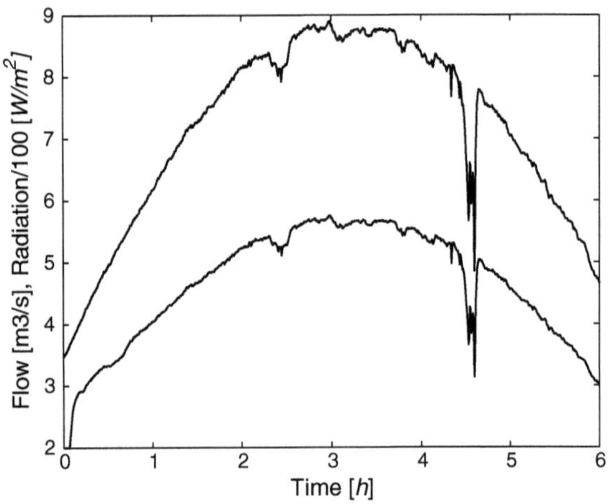

Fig. 6.12 Example 6.1. Radiation ($\times 10^{-2}$ Wm^{-2} —*upper curve*) and flow ($\times 10^4$ l/s—*lower curve*)

that are lower for lower coordinates of the collocation point (in other words, in equilibrium, at given time instant, the temperature decreases along space, from the beginning to the end of the DCSF pipe).

Figure 6.11 shows the square of the tracking error that, as expected, converges to zero. Finally, Fig. 6.12 shows the solar radiation and the physical manipulated variable, given by the fluid flow. The solar radiation is the main disturbance acting on the system, and is measured to compute the physical manipulated variable in the linearizing feedback compensator. With no surprise, the fluid flow is much influenced by the solar radiation.

Example 6.2 In example 6.2, two sets of simulations concerning the adaptive RHC are reported, with different values of the horizon T and N_u selected accordingly. In

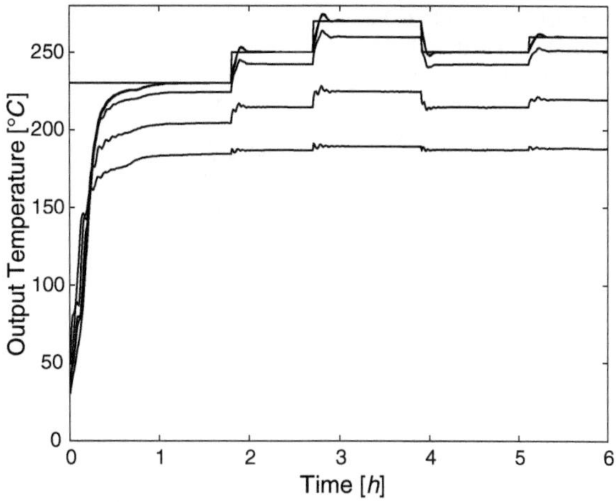

Fig. 6.13 Example 6.2(a). Closed loop with RHC having $T = 60$ s. Outlet fluid temperature and reference and temperature estimates at the collocation points

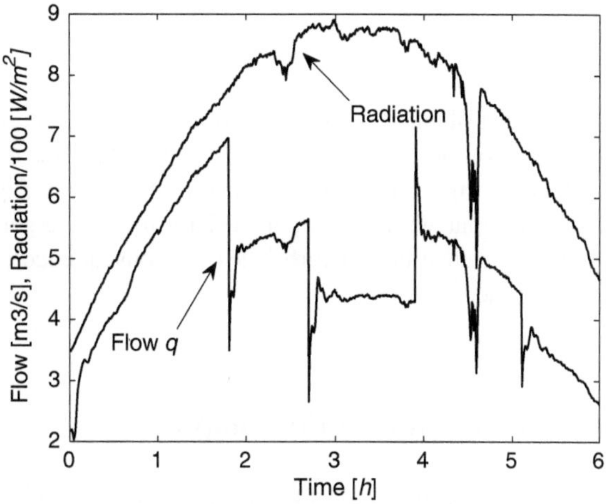

Fig. 6.14 Example 6.2(a). Closed loop with RHC having $T = 60$ s. Radiation (disturbance—*above*) $(\times 1 \times 10^{-2})$ [W/m^2] and fluid flow (manipulated variable — *below*) $(\times 1 \times 10^4)$ l/s

the first $T = 60$ s $N_u = 2$, while in the second $T = 180$ s $N_u = 6$. Here, N_u denotes the number of time "strips" in which the prediction horizon is decomposed when approximating the integral (6.39) by Rieman sums, for the numerical computation of the control sequence.

Figures 6.13 through 6.15 show the results for the controller configuration with $T = 60$ s, while Figs. 6.16 through 6.18 show the results for the configuration with $T = 180$ s. Apart from the value of N_u, the major difference between these two controller configurations is the prediction horizon T which is larger in the second case. As expected, a larger horizon yields smoother input and output signals. Comparing Fig. 6.13 (for $T = 60$) with Fig. 6.16 (for $T = 180$) it is seen that, by increasing the horizon, the output response becomes less oscillatory. This fact is also true for the manipulated variable (compare Fig. 6.14, obtained for $T = 60$, with Fig. 6.17, obtained for $T = 180$).

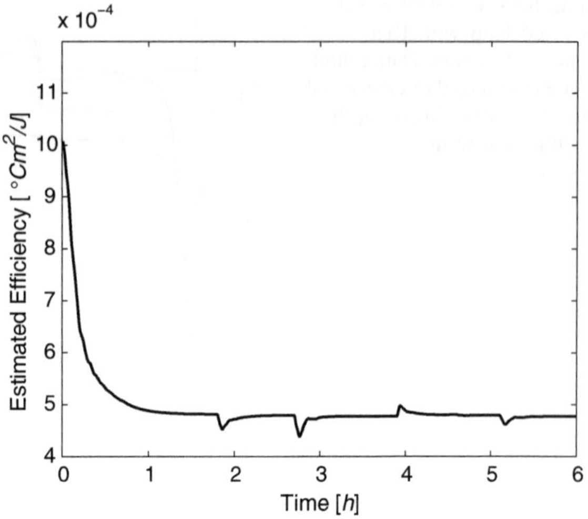

Fig. 6.15 Example 6.2(a). Closed loop with RHC having $T = 60\,\text{s}$. Mirror efficiency estimate, $\hat{\alpha}$

The test planned includes a strong and fast disturbance at time 4.5 h, as seen in the radiation plots of both Figs. 6.14 and 6.17. This disturbance is well rejected.

Comparing the estimates of mirror efficiency α, Fig. 6.15 ($T = 60$) and Fig. 6.18 it is seen that with the larger horizon the convergence is slightly slower, but the reference transitions disturb it less, a consequence of the improvement in control performance.

6.4 Main Points of the Chapter

In this chapter, the control has been design by combining three elements:

- A change of variables that exactly linearizes the system dynamics;
- A parameter estimator;
- A control law for the exactly linearized system.

While these elements are also present in Chap. 5, the technique used to implement them is much different here, in what concerns the first two. Indeed, while in Chap. 5 the variable used to linearize the system is time, in the present chapter it is the manipulated input. Furthermore, while Chap. 5 relies on an a priori defined parameter estimator (recursive least squares are used), without any assurance on stability, in the present chapter the estimation rule results from a condition that ensures stability of the overall system. For that sake a joint Lyapunov function for control and estimation is used.

The control law used for the exactly linearized system that results from the combination of the first two procedures can be either a state feedback controller or a model predictive controller that minimizes a receding horizon cost. The use of MPC is computationally much more demanding, a limitation that is not critical given the low

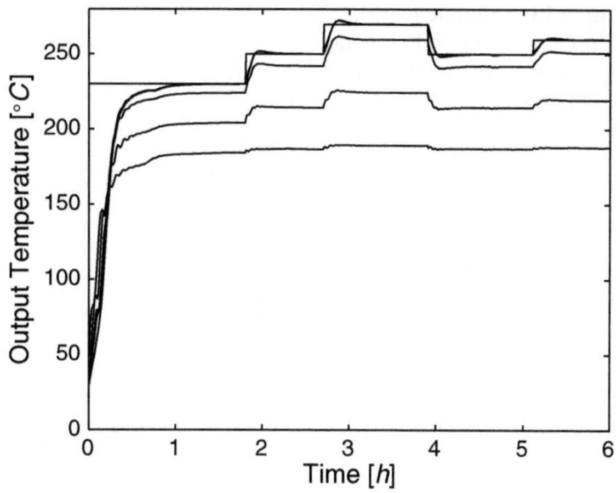

Fig. 6.16 Example 6.2(b). Closed loop with RHC having $T = 180$ s. Outlet fluid temperature and reference and temperature estimates at the collocation points

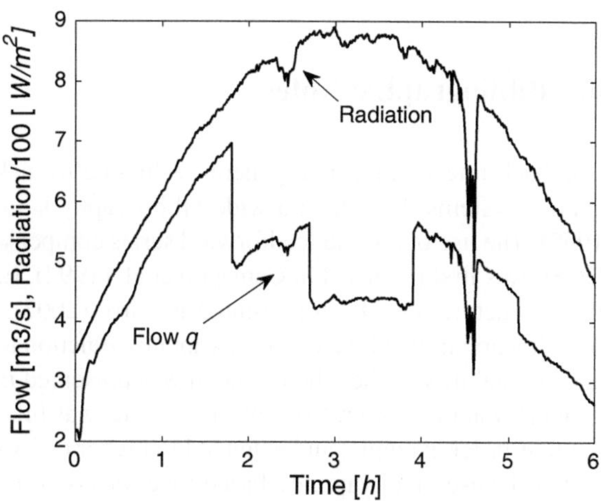

Fig. 6.17 Example 6.2(b). Closed loop with RHC having $T = 180$ s. Radiation (disturbance—*above*) $(\times 1 \times 10^{-2})$ [W/m^2] and fluid flow (manipulated variable —*below*) $(\times 1 \times 10^4)$ l/s

sampling rates used in DCSF control, but has the significant advantage of allowing to incorporate constraints in manipulated variable computation.

When using MPC, the constraints can be of two types: Operational and stability. Operational constraints refer to the allowable range of variables (for instante, minimum and maximum values of the fluid flow, or of the fluid temperature). Stability constraints refer to constraints that are imposed in order to ensure stability of the closed-loop system with the resulting controller. In this chapter, a stability constraint is imposed on the cost value that results from a controller that ensures stability by minimizing a Lyapunov function. This other controller forms a kind of "safety net" that ensures a stable solution to the control problem from which other solutions with increased performance can be searched.

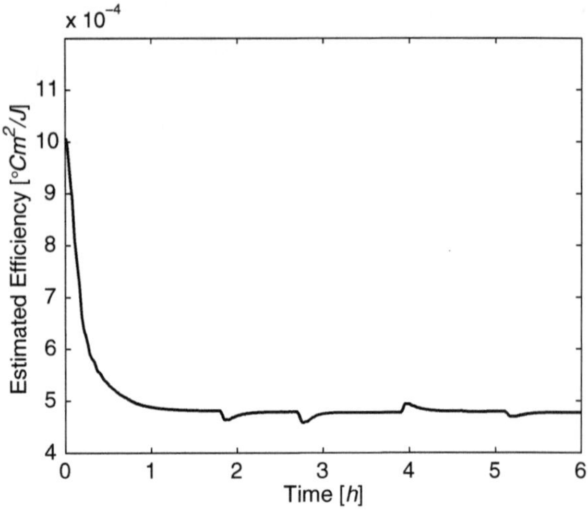

Fig. 6.18 Example 6.2(b). Closed loop with RHC having $T = 180\,\text{s}$. Mirror efficiency estimate, $\hat{\alpha}$

6.5 Bibliographic Notes

Feedback linearization is a general technique to tackle some classes of nonlinear control systems, for which a wide bibliography is available (Slotine 1991; Isidori 1995). The use of a static feedforward series compensator for the nonlinearity of the DCSF was first proposed in Camacho et al. (1992). Exploitations and extensions of these structure have been presented in Cirre (2007). The idea of combining feedback linearization with a control Lyapunov function to obtain an adaptation law that ensures stability of the whole system was presented in Sastry and Isidori (1989) for general systems. For DCSF control, this idea was first presented in Barão (2002) for the first-order example, and extended to higher order models in Igreja et al. (2003).

For the use of a Lyapunov-based function controller to impose a stability bound, see Sontag (1998), Primbs (2000).

References

Barão M, Lemos JM, Silva RN (2002) Reduced complexity adaptive nonlinear control of a distributed collector solar field. J Proc Control 12:131–141

Camacho EF, Rubio FR, Hughes FM (1992) Self-tuning control of a solar power plant with a distributed collector field. IEEE Control Syst Mag 12(2:)72–78

Cirre CC, Berenguel M, Valenzuela L, Camacho EF (2007) Feedback linearization control for a distributed solar collector field. Control Eng Pract 15:1533–1544

Findeisen R, Allgöwer F (2002) An introduction to nonlinear model predictive control. In : Proceedings of 21st Benelux meeting on systems and control, Veldhoven

Igreja, JM, Lemos JM, Barão M, Silva RN (2003). Adaptive nonlinear control of a distributed collector solar field. In: Provceedings of European control conference 2003, ECC03, Cambridge UK

Isidori, A (1995) Nonlinear control systems. Springer, New York

Primbs JA (2000) A receding horizon generalization of pointwise min-norm controllers. IEEE Trans Autom Control 45:898–909

Sastry S, Isidori A (1989) Adaptive control of linearizable systems. IEEE Trans Autom Control 34(11):1123–1131

Slotine, J (1991) Applied nonlinear control. Prentice Hall, NJ

Sontag, E (1989) Mathematical control theory, 2nd edn. Springer, New York

Chapter 7
Adaptive Motion Planning

There are many control problems in which the objective consists of tracking a time-varying reference that is known in advance. The problem can be formulated, e.g., as a predictive or an optimal control problem (Lewis and Syrmos 1995; Mosca 1995), either in the time or frequency domains, but it always involves, in a way or another, a model of the signal to track and leads to a solution in the form of a two degrees of freedom controller structure (Mosca 1995). While in some situations the profile of the manipulated variable is a priori unknown, there are many situations in which an input nominal profile to be followed by the process output and/or the state is specified in advance. In this last case, advantage may be taken of this knowledge in order to design a servo controller, i.e., a controller that tracks a variable reference, such as to improve its performance.

The problem is then twofold: First, to compute nominal input and output profiles, possibly using an approximated plant model, and then to design a regulator that stabilizes the actual plant trajectory around them. For finite dimensional linear controllable systems the solution to these problems is well known (Henson and Seborg 1997).

In other, more complicated situations, such as the ones involving nonlinear and/or distributed parameter systems, the concept of flatness (Fliess et al. 1995) is instrumental in designing the motion planner that yields the nominal trajectory and in coupling it with the deviations regulator (Mounier and Rudolph 1998). Flatness generalizes to nonlinear systems the notion of controllability. Indeed, for flat systems, it is possible to define a virtual output, known as a *flat output* such that all the process inputs and states can be expressed as an explicit function of this flat output and its derivatives. In turn, this function can be used to invert the system dynamics and compute the manipulated variable evolution in time that steers the process output and the state from one value to another, along a prescribed path. For DCSF models, one must resort to the concept of orbital flatness, that involves a change of the time scale as done on Chap. 5 and that will be considered again extensively hereafter.

J. M. Lemos et al., *Adaptive Control of Solar Energy Collector Systems*,
Advances in Industrial Control, DOI: 10.1007/978-3-319-06853-4_7,

This chapter is concerned with the design of an adaptive servo controller for solving the motion problem, i.e., for tracking variable references in a DCSF that are planned in advance. The structure proposed is made up of three main blocks:

- A motion planner;
- An incremental controller;
- An adaptation mechanism.

The motion planner selects the time profile of the manipulated variable (fluid flow) such that the plant state (fluid temperature distribution along the solar field) is driven between successive equilibrium states as specified. This is done on the basis of a simplified, yet distributed parameter, model and uses the methods of flat systems and the concept of orbital flatness. In order to stabilize the actual fluid temperature around this nominal path, a linear controller is used. This control law is then modified according to a Lyapunov function strategy for incorporating adaptation through the adjustment of a parameter that conveys the most significative plant uncertainty.

For the sake of easy reference, the dominant dynamics of the field discussed in Chap. 2 is rewritten again hereafter by

$$\frac{\partial T(z,t)}{\partial t} + u(t)\frac{\partial T(z,t)}{\partial z} = \alpha R(t), \tag{7.1}$$

where $T(z,t)$ denotes the fluid temperature at position z and at time t, u is the fluid velocity (proportional to flow), taken as the manipulated variable, R is solar radiation, assumed to be accessible for measure, and α is a parameter that is assumed to be constant or, at least slowly varying in the time scale of the DCSF state. In the initial stage of the design, the parameter α is assumed to be known. In reality, this parameter is unknown since it depends on the fluid-specific heat, that in turn varies with the temperature, and on the mirror reflectivity that, as discussed in Chap. 1, may be affected by unpredictable factors (e.g., mirror deformation due to the wind or dust deposition). To cope with this lack of knowledge, the true value of α is replaced, foir the purposes of controller design, by its estimated $\hat{\alpha}$, issued by an adaptation algorithm. The length of the pipe is denoted by L. The state of this distributed parameter system is described at each time t by the function $\{T(z,t),\ 0 \leq z \leq L\}$.

The chapter is organized as follows: First, motion planning using the concept of orbital flatness is discussed for the DCSF infinite-dimensional model given by (7.1). Then, the same procedure is followed for the moisture control system presented in Sect. 2.5.1.

7.1 Flatness-Based Motion Planning

In simple terms, a system is said to be flat (Fliess et al. 1995) if it is possible to find a set of outputs, called the flat outputs, that are equal in number to the number of inputs, and such that all the states and inputs can be determined from these outputs without integration.

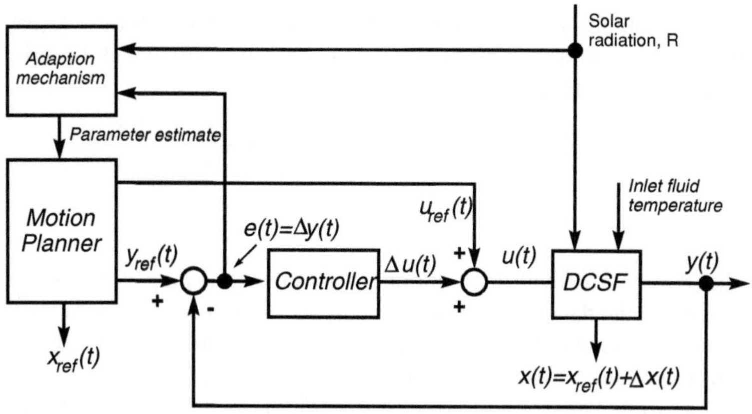

Fig. 7.1 Motion planning-based servo control structure

If the system variables are to follow a given time profile, this objective of tracking a reference can be specified in terms of the flat outputs, and the above functions are then used to compute the corresponding inputs. This formulation is, in rough terms, the motion planning problem (Lynch and Rudolph 2002). For SISO systems, the planning performed in the above way is equivalent to have an "artificial" output through which it is possible to perform exact feedback linearization.

Figure 7.1 shows the block diagram of a general motion planning-based servo control structure (Mounier and Rudolph 1998). The motion planner block computes the nominal manipulated value u_{ref} that, in the ideal case where there are no disturbances and the model perfectly matches the plant, leads the process output y to be equal to the desired reference y_{ref}. The feedback controller compensates only for deviations in relation to the planned trajectory $(x_{\text{ref}}(t),\ u_{\text{ref}}(t))$ due to disturbances and uncertainties. Finally, the adaptation block provides an estimate of the uncertain parameter to the other blocks in order to tackle parameter uncertainty.

7.1.1 Orbital Flatness

Although for the DCSF model (7.1) a flat output is not known, it is possible to introduce a *time scaling* such that the system becomes flat (Fliess et al. 1995; Guay 1999; Respondek 1998). Such a system is then said to be orbitally flat. Thus, the solution to the problem at hand is obtained by introducing a change of the time variable $\tau(t)$ such that, in the new coordinate space $(z,\ \tau)$ a flat output $\bar{y}$ is given by:

$$\hat{y}(\tau) = h(y(z,\ \tau),\ u,\ \dot{u},\ \ldots), \tag{7.2}$$

where h is a function to be found.

Once the flat output is found, it is possible to describe all trajectories $y(z, \tau), \ u(\tau)$ that satisfy the transformed PDE as a function of the flat output and its derivatives, namely, there are functions φ and χ such that

$$y(z, \tau) = \varphi(\hat{y}, \dot{\hat{y}}, \ddot{\hat{y}}, \ldots), \tag{7.3}$$

$$u(\tau) = \chi(\hat{y}, \dot{\hat{y}}, \ddot{\hat{y}}, \ldots). \tag{7.4}$$

Equations (7.3), (7.4) provide the solution to the motion planning problem. Indeed, if a given shape for the flat output $\bar{y}$ is imposed, (7.4) yields the input profile that drives the state $y(z, \tau)$ as required. The use of this technique to perform motion planning for the DCSF, assumed to be described by (7.1), is explained hereafter.

According to the approach of orbital flatness, consider the change of time scale already introduced in Chap. 5, where τ is a new time scale that is related to the "real" time t by

$$\tau(t) = \int_0^t u(\sigma) d\sigma \tag{7.5}$$

or, equivalently,

$$\frac{\mathrm{d}\tau}{\mathrm{d}t} = u(t), \tag{7.6}$$

where u is the manipulated variable in the DCSF model (7.1), given by the fluid velocity. As discussed in Chap. 5 , this change of variable introduces a "natural" time scale associated to fluid flow that linearizes the plant model, forcing the characteristic lines of (7.1) to become straight lines.

It is assumed that the mapping between t and τ is bijective, that τ is a monotonically function of t, and that τ goes to infinity if and only if t goes to infinity. The validity of these assumptions is ensured by natural physical constraints in the practical problem at hand. Indeed, $u(t)$ is constrained to be positive and belongs to the interval $[u_{\min}, u_{\max}]$. Under these hypothesis, the "real" time t can be recovered from the transformed time τ from

$$t(\tau) = \int_0^\tau \frac{1}{u(\sigma)} d\sigma. \tag{7.7}$$

Since, using the chain rule for derivatives

$$\frac{\partial T}{\partial t} = \frac{\partial T}{\partial \tau}\frac{\partial \tau}{\partial t} = \frac{\partial T}{\partial \tau}u, \tag{7.8}$$

in the time scale τ, equation (7.1) becomes

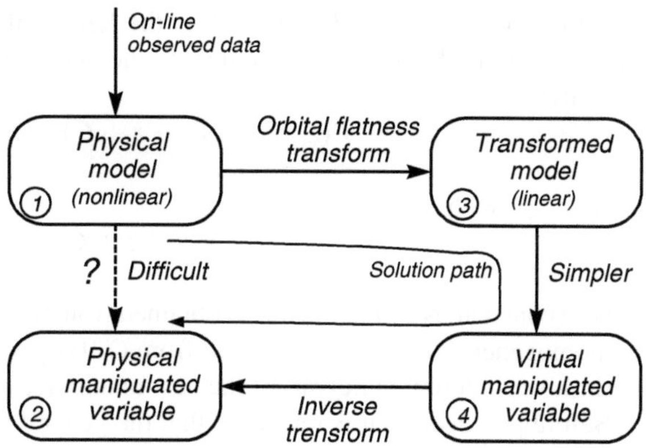

Fig. 7.2 Solving the motion planning problem for a DCSF using orbital flatness

$$\frac{\partial T(z,\tau)}{\partial \tau} + \frac{\partial T(z,\tau)}{\partial z} = f(\tau), \tag{7.9}$$

where

$$f(\tau) \triangleq \frac{\alpha R(t(\tau))}{u(t(\tau))} \tag{7.10}$$

is the new manipulated variable in the transformed coordinates. This manipulated variable is also called "virtual" because it has only mathematical existence, although its computation is an important step to compute the physical manipulated variable u that is actually applied to the DCSF.

The use of the model (7.9) with the transformed variables to solve the motion planning problem associated to a DCSF represented by (7.1) is explained on the diagram of Fig. 7.2. The direct solution of this problem to get a physical manipulated variable (i.e., a "real" time profile for the fluid flow) that drives the plant state along the desired path of temperatures is difficult because the DCSF model (7.1) includes an algebraic product between the manipulated variable u and the partial derivative of the fluid temperature with respect to space. In the diagram of Fig. 7.2, this solution corresponds of moving from block 1 to block 2.

In order to circumvent the above difficulty, the orbital flatness change of variable (7.5) is performed, transforming the original DCSF model to (7.9). In terms of Fig. 7.2, this transformation corresponds to moving from block 1 to block 3.

Since with this transform the model becomes linear, it is now much simpler to solve the motion planning problem so as to get the desired virtual manipulated variable f or, in terms of Fig. 7.2, to move from block 3 to block 4.

In order to obtain the physical manipulated variable, the expression (7.10) ought to be inverted and combined with the inverse change of the time variable to obtain $u(t)$ as a function of $f(\tau)$. In terms of Fig. 7.2, this step is the final one, that leads from block 4 to the desired answer provided by block 2.

In order to implement the above procedure, we start by considering the general solution of the transform model (7.9). For this sake, observe that, with a boundary condition given for $z = L$, where L is the length of the pipe and z is the length measured along the pipe, from the inlet to the outlet, the general solution of (7.9) is given by:

$$T(z,\tau) = \phi(\tau - z + C) + F(z), \tag{7.11}$$

where $\phi(\xi)$, with

$$\xi = \tau - z + C \tag{7.12}$$

and C a constant, is a function that is obtained from the initial conditions that satisfies the homogeneous equations, obtained from (7.9) by making $f(\tau) = 0$, and $F(\tau)$ is a primitive (indefinite integral) of the new manipulated variable $f(\tau)$.

Before proceeding, it is remarked that the general solution (7.11) is obtained by imposing a boundary condition for $z = L$ and corresponds to a wave that propagates in time from the pipe outlet to the pipe inlet. Instead, the general solution (2.6) is obtained with a boundary condition imposed for $z = 0$ and corresponds to a wave that propagates in the opposite direction.

Let $\bar{y}(\tau)$ be a known given function of τ and consider the boundary condition, relative to the outlet fluid temperature gradient with respect to z, computed for $z = L$ and given by:

$$\left.\frac{\partial T(z,\tau)}{\partial z}\right|_{z=L} = \bar{y}(\tau). \tag{7.13}$$

Imposing the boundary condition (7.13) and considering the definition of ξ given by (7.12) implies that

$$\bar{y}(\tau) = \left.\frac{\partial T(z,\tau)}{\partial z}\right|_{z=L} = \left.\frac{d\phi(\xi)}{dx}\right|_{z=L} \frac{\partial \xi}{\partial z} = -\left.\frac{d\phi(\xi)}{d\xi}\right|_{z=L}. \tag{7.14}$$

Furthermore, by making $z = L = C$, the total derivative of ϕ with respect to ξ is equal to the derivative with respect to τ (because with z constant, ξ depends only on τ), and (7.14) implies that

$$\frac{d\phi(\tau)}{d\tau} = -\bar{y}(\tau) \tag{7.15}$$

or

$$\phi(\tau) = -\bar{Y}(\tau), \tag{7.16}$$

where $\bar{Y}(\tau)$ is a primitive function of $\bar{y}(\tau)$. Replacing this expression of ϕ on the general solution (7.11) yields

$$T(z,\tau) = F(\tau) - \bar{Y}(\tau + L - z). \tag{7.17}$$

The manipulated function $f(\tau)$ is obtained from the boundary condition for $z = 0$, yielding

$$T(0,\tau) = F(\tau) - \bar{Y}(\tau + L). \tag{7.18}$$

Assuming the inlet fluid temperature to be constant, differentiation of (7.18) with respect to τ implies that

$$f(\tau) = \bar{y}(\tau + L), \tag{7.19}$$

and

$$T(z, \tau) = T_0 + \bar{Y}(\tau + L) - \bar{Y}(\tau + L - z), \tag{7.20}$$

where T_0 is the inlet fluid temperature, that is assumed to be constant.

Expanding $\bar{y}(\tau + L)$, $\bar{Y}(\tau + L)$, and $\bar{Y}(\tau + L - z)$ in a Taylor series around τ we get

$$f(\tau) = \sum_{k=0}^{\infty} \frac{L^k}{k!} \overset{(k)}{\bar{y}}(\tau) \tag{7.21}$$

and

$$T(z, \tau) = T_0 + \sum_{k=1}^{\infty} \frac{L^k}{k!} \overset{(k-1)}{\bar{y}}(\tau) - \sum_{k=1}^{\infty} \frac{(L-z)^k}{k!} \overset{(k-1)}{\bar{y}}(\tau). \tag{7.22}$$

The above expressions provide the algebraic expressions needed for dynamic motion planning. The trajectories are expressed as algebraic functions of the inlet fluid temperature T_0 (assumed to be constant) and of the successive derivatives with respect to time of the flat output $\bar{y}$. Planning is possible as long as the series converge.

7.1.2 Trajectory Generation

Motion planning connects stationary states, for which the fluid temperature along the pipe is given by a function $T_{ss}(z)$ that is independent from time. Hence, the partial derivative with respect to time in (7.9) vanishes for the steady-state distribution and it follows that

$$\frac{dT_{ss}(z)}{dz} = f_{ss}, \tag{7.23}$$

where f_{ss} is the equilibrium value of the manipulated variable in the modified co-ordinates, that is equal to the gradient of the temperature with respect to space, and hence

$$T_{ss}(z) = f_{ss} z + T_0. \tag{7.24}$$

Planning is made such that the temperature distribution along the pipe moves from the stationary state

$$T(z, 0) = C_1 z + C_{01}, \tag{7.25}$$

with

$$T_0 = C_{01} \tag{7.26}$$

and

$$T(L, 0) = C_1 L + C_{01}, \tag{7.27}$$

to the new stationary state, to be attained at transformed time τ^*

$$T(L, \tau^*) = C_2 L + C_{01}. \tag{7.28}$$

An immediate consequence is that the constants C_1 and C_2 are computed from the boundary conditions by:

$$C_1 = \frac{T(L, 0) - T_0}{L} \tag{7.29}$$

and

$$C_2 = \frac{T(L, \tau^*) - T_0}{L}. \tag{7.30}$$

The trajectory generator block produces a reference signal and the corresponding derivatives with respect to time up to a generic order n. Each trajectory is made of pieces that correspond to functions that pass by two consecutive way points. This means that, once the $K + 1$ way points $P_0, P_1, \ldots, P_K$ defined, the trajectory is generated between P_0 and P_1, P_1 and $P_2, \ldots,$ up to P_{K-1} and P_K. The transition time between two consecutive way points is such that the trajectory passes by $P_0(0), P_0(t_1), \ldots, P_K(t_K)$, that is to say, passes by the way points and at definite transit times.

The trajectory connecting two stationary states at times $\tau = 0$ and $\tau = \tau^*$ is such that its derivatives at the beginning and at the end of the interval in which they are defined are zero. This imposition prevents the use of analytic functions because these would vanish in the entire interval. Hence, a particular type of functions, known as exponential type Gevrey function of class α (Rudolph et al. 2003) is used.

Therefore, the profile for changing the flat output is given by:

$$y(\tau) = C_1 + (C_2 - C_1)\Phi_{y\sigma}\left(\frac{\tau}{\tau^*}\right), \tag{7.31}$$

where C_1 and C_2 are given, respectively, by (7.29) and (7.30), and $\Phi_{y\sigma}$ is an exponential type Gevrey function of class α, further discussed below, that verifies

$$\Phi_{y\sigma}(0) = 0 \quad \tau \leq 0, \tag{7.32}$$

$$\Phi_{y\sigma}(1) = 1 \quad \tau \geq \tau^*. \tag{7.33}$$

7.1.3 Exponential Type Gevrey Functions of Class α

A Gevrey function of class α (Rudolph et al. 2003) is a function $\Phi_\sigma : [0, \tau^*] \to \mathbb{R}$, the derivatives of which are bounded on the interval $[0, \tau^*]$ by

$$\sup_{t\in[0,\tau^*]} |\Phi_\sigma^{(k)}(\tau)| \leq m\frac{(k!)^\alpha}{\gamma^k}. \tag{7.34}$$

Hereafter, we use exponential type Gevrey functions of class α that are defined by:

$$\Phi_\sigma(t) = \begin{cases} \frac{\int_0^{\frac{t}{t^*}} \phi_\sigma(\tau)d\tau}{\int_0^1 \phi_\sigma(\tau)d\tau} & 0 \leq t \leq t^* \\ 1 & t > t^* \end{cases}, \tag{7.35}$$

where

$$\phi_\sigma(t) = \begin{cases} e^{-\frac{1}{(\tau(1-\tau))^\sigma}} & 0 \leq t \leq 1, \quad \tau = \frac{t}{t^*} \\ 0 & \tau > 1 \end{cases}. \tag{7.36}$$

The function defined by (7.35) varies monotonously between C_i and C_{i+1}. The parameter σ imposes the rate of change of the transition. The higher the value of σ, the faster is the transition. The derivatives of any order, computed at $t = 0$ and $t = 1$ are zero. It is also remarked that $\Phi_\sigma : \mathbb{R}^+ \to \mathbb{R}$ is a nonanalytical function that belongs to $\mathcal{C}^\infty$ and that $y(t)$ is also an exponential Gevrey function of class α.

7.1.4 Computing the Output Derivatives

Numerical differentiation of the flat output does not provide acceptable results. Instead, the derivative of order n of $y(t)$ can be computed from the derivative of order $n-1$ of $\phi_\sigma(\tau)$ using the recursive formulas

$$\frac{\mathrm{d}^n y(t)}{\mathrm{d}t^n} = (C_{i+1} - C_i)\frac{\mathrm{d}^n \Phi_\sigma(t)}{\mathrm{d}t^n} \tag{7.37}$$

$$\frac{\mathrm{d}^n \Phi_\sigma(t)}{\mathrm{d}t^n} = \frac{1}{\int_0^1 \phi_\sigma(\zeta)\mathrm{d}\zeta} \frac{\mathrm{d}^n (\int_0^\tau \phi_\sigma(\zeta)\mathrm{d}\zeta)}{\mathrm{d}t^n} \tag{7.38}$$

$$\frac{\mathrm{d}^n \Phi_\sigma(t)}{\mathrm{d}t^n} = \frac{1}{\int_0^1 \phi_\sigma(\zeta)\mathrm{d}\zeta} \frac{1}{t^{*n}} \frac{\mathrm{d}^{n-1}\phi_\sigma(\tau)}{\mathrm{d}\tau^{n-1}} \tag{7.39}$$

7.1.5 Computation of the Derivatives of a Gevrey Function

Recalling (7.35), inside the transition interval where the Gevrey function is defined, it happens that

$$\phi_\sigma(t) = e^{u(t)} = e^{-\frac{1}{(t(1-t))^\sigma}}. \tag{7.40}$$

Defining $f(t) = e^{u(t)}$, the following recursive formula can be obtained by successive differentiation

$$\overset{(n)}{f} = \sum_{k=1}^{n} \binom{n-1}{k-1} \overset{(k)}{u} \overset{(n-k)}{f} , \qquad n \geq 1. \tag{7.41}$$

In order to get the derivative of order n of $u(t)$ it is enough to observe that

$$\dot{u} = \frac{\sigma(2t-1)}{t(1-t)} u \tag{7.42}$$

or

$$\begin{aligned} \dot{u} u_1 &= \dot{u}_2 u, \\ u_1 &= t(1-t), \\ \dot{u}_2 &= \sigma(2t-1). \end{aligned}$$

Upon successive differentiation of this equality, the following generic recursive formula is obtained

$$u_1 \overset{(n+1)}{u} + [n\dot{u}_1 - \dot{u}_2] \overset{(n)}{u} + [n(n-1)\ddot{u}_1 - n\ddot{u}_2] \overset{(n-1)}{u} = 0 , \qquad n \geq 2 \tag{7.43}$$

with

$$\dot{u} = \frac{\sigma(2t-1)}{t(1-t)} u \tag{7.44}$$

and

$$\ddot{u} = \frac{\sigma(2t-1)}{t(1-t)} \dot{u} + \left[\frac{(1-2t)^2}{t(1-t)^2} + \frac{2}{t(1-t)} \right] \sigma u. \tag{7.45}$$

The use of the above recursive formulas allows to obtain in an efficient way the derivatives of $y(t)$ up to order n, without requiring numerical differentiation where the error would be very difficult to control.

7.1.6 Example: Motion Planning of a DCSF

Figures 7.3, 7.4, and 7.5 illustrate the trajectories generated for the simulated situation with $L = 180$, $\alpha R(t) = 1$ and $T_0 = 0$, and in which three successive transitions take place between equilibrium states. As previously explained, the planning is initially performed in transformed time τ. Figure 7.3 shows the temperature gradient, $y(\tau)$, and the corresponding input, $f(\tau)$, in transformed time. Once the planning is complete in transformed time, real time is recovered using Eq. (7.7). The relation between transformed time and real is shown in Fig. 7.4. The planning is then referred

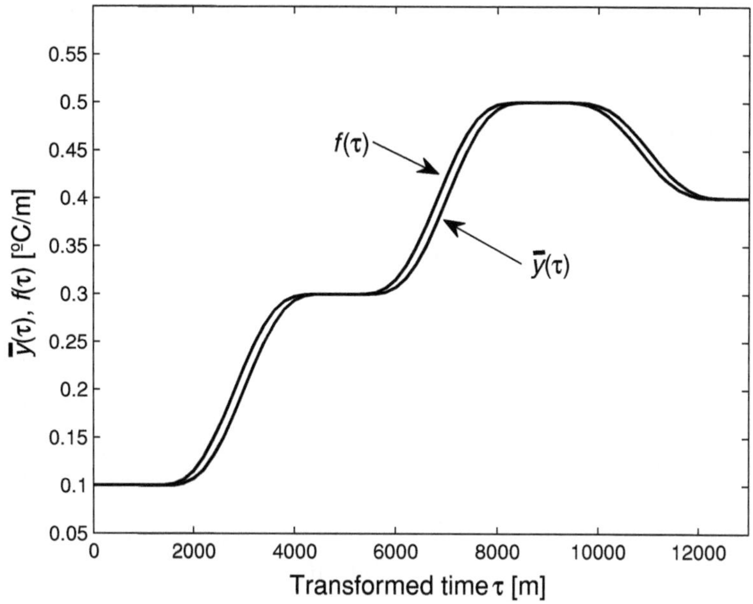

Fig. 7.3 Temperature gradient, $y(\tau)$, and the corresponding input, $f(\tau)$, in transformed time

to real time by using the relation expressed in Fig. 7.4 to transform the time scale and get Fig. 7.5.

Of course, after obtaining $t(t)$ from $f(\tau)$ and the change of variable (7.7) (depicted graphically in Fig. 7.4) it is possible to compute the fluid velocity $u(t)$, that corresponds to the actual physical manipulated variable to be applied to the DCSF.

As can be seen in Fig. 7.5, at each time instant the fluid temperature distribution along the pipe is an increasing function of the pipe length, with its minimum at the value of the inlet fluid temperature (for $z = 0$) and increasing along the pipe, to reach its maximum at the outlet. The fact that the temperature cannot decrease along the pipe is a fundamental constraint imposed by the fact that there is not a cooling mechanism. However, by changing the fluid velocity (or, equivalently, its flow, since the cross section of the pipe is constant and the fluid is incompressible), it is possible to change the heating rate of the fluid (since, as discussed in Chap. 2, the highest the fluid velocity, the smallest the energy accumulated in each fluid element during its transit time) and hence the temperature at the pipe outlet. In this way, it is possible to manipulate the outlet fluid temperature by increasing and decreasing its value, as shown in Fig. 7.5.

It is also remarked that, since the units of u are m/s (because u is a velocity), from (7.6) it is concluded that the units of the transformed time τ are units of distance. This fact is not surprising since, as discussed in Chap. 5, the change of time variable (7.6) renders the characteristic lines of (7.9) straight in the space $[z, \tau]$. Therefore, in the transformed time τ, the solution of (7.9) is such that it is constant along lines

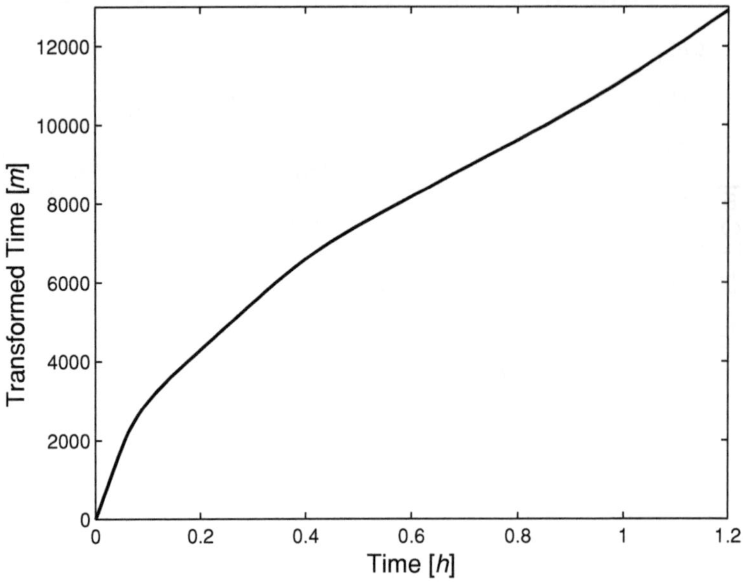

Fig. 7.4 Recovering real-time $t(\tau)$ from transformed time τ

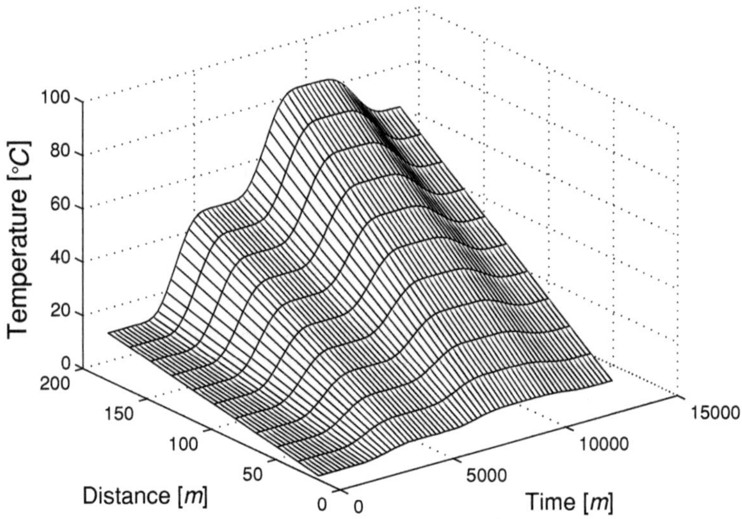

Fig. 7.5 Temperature evolution in real time $T(z, t)$

in $[z, \tau]$ such that time is proportional to space. In physical terms (and repeating an argument already discussed in Chap. 5), this means that, when a fluid segment of length Δz progresses along the pipe of a length Δz, in transformed time it always takes the same time interval to do this progression.

7.2 Servo Control

Consider now the design of the block named "controller" in Fig. 7.1. The problem of tracking a reference trajectory for flat SISO finite-dimensional systems can be formulated as follows: (Henson and Seborg 1997): Given a nonlinear system of finite dimension n and state x described by:

$$\dot{x} = F(x, u), \tag{7.46}$$

with flat output

$$\hat{y} = h(x), \tag{7.47}$$

find a controller such that, if x_r is the state to track than it generates the corresponding manipulated variable u_r that, if applied to the plant, yields x_r.

If the dynamics admits a flat output, it is possible, by a change of variable, to transform the system to the normal form

$$\overset{(n+1)}{\bar{y}} = \nu, \tag{7.48}$$

where ν is a transformed (also called "virtual") manipulated variable. By computing ν according to the control law

$$\nu = \overset{(n+1)}{\hat{y}}_r - K\Delta\tilde{y}, \tag{7.49}$$

where

$$\bar{y}_r(t) = h(x_r(t)),$$

$$\tilde{y}(t) = \left[\, \bar{y}(t)\ \dot{\bar{y}}(t)\ \ldots\ \overset{(n)}{\bar{y}}(t) \,\right]^T$$

$$K = \left[\, k_1\ k_2\ \ldots\ k_n \,\right]^T,$$

and

$$\Delta y = y - y_r,$$

is the tracking error, the closed-loop error satisfies

$$\overset{(n+1)}{\Delta\hat{y}} = -K\Delta\tilde{y}. \tag{7.50}$$

It is possible to select the vector of gains K such that the closed-loop dynamics (7.50) becomes asymptotically stable. Tracking is therefore reduced to the closed-loop control of a series of n integrators whose error dynamics is linear and given by (7.50). Figure 7.1 shows a general block diagram for this servo controller. By selecting the gains of the vector K in a suitable way, it is possible to impose a stable dynamics for the closed-loop tracking error.

In the special case of DCSFs, the main difficulty consists of the fact that it is only possible to write the system equations in normal form in the domain of the transformed time τ. Other difficulties arise when the system is of distributed parameter type (meaning that it is of infinite dimension and described by ordinary differential equations), when the output to control is not coincident with the (orbital) flat output, and due to the technological limitation that temperature is only measured at the outlet and not inside the pipe.

Under these circumstances, one possibility to solve the regulation problem (i.e., computing incremental corrections to the manipulated variable such that the process output remains close to the reference issued by the motion planner) might be to use the time scale (7.7), together with the state feedback control law (7.49).

As shown above, it is possible to obtain trajectories for fluid flow (or velocity) and for the outlet fluid temperature, denoted, respectively, $u_r(t)$ and $T_{or}(t)$. Having access to these signals, another option is thus to close the loop through a PID controller that produces a compensation of the deviations with respect to the reference trajectory $(T_{or}(t),\ u_r(t))$. The compensation feedback controller can be designed using, e.g., the methods of Chap. 3 in order to obtain variations of either the GPC or of the MUSMAR algorithms that minimize, in a receding horizon sense, the quadratic cost

$$J_\Delta(k) = \mathcal{E}\left\{\sum_{i=1}^{T}(T_{out}(k+i) - T_{or}(k+i))^2 + \rho(u(k+i-1) - u_r(k+i-1))^2 \,|\mathcal{O}^k\right\}, \tag{7.51}$$

where T_{out} is the outlet fluid temperature and k is the current discrete time.

Another possibility, obtained *ad hoc* but that proved to yield adequate results, is to tune the parameters K_C, T_I and T_D of a PID controller that yields the incremental correction Δu of the manipulated variable according to

$$\Delta u(t) = -K_C\left(e(t) + \frac{1}{\tau_I}\int_0^t e(\zeta)\mathrm{d}\zeta + \tau_D\frac{\mathrm{d}e(t)}{\mathrm{d}t}\right), \tag{7.52}$$

where the tracking error is

$$e(t) = \Delta T(t) = T_{or}(t) - T_{out}(t). \tag{7.53}$$

The manipulated variable actually applied to the DCSF is then computed by

$$u(t) = u_r(t) + \Delta u(t). \tag{7.54}$$

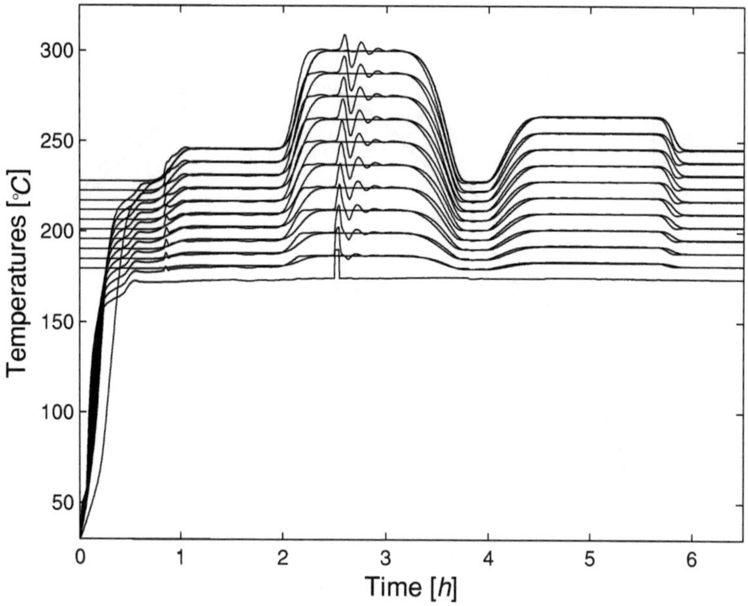

Fig. 7.6 Planning for the solar field: Temperatures along the pipe and their references [°C]

Of course, the basic PID algorithm defined by (7.52) should be modified to include the "fixes" required by practical situations, including antireset of the integrator and the replacement of the pure derivative action by a high-pass filter.

The above scheme assumes that a nominal trajectory for the sun radiation is known in advance and that the inlet fluid temperature is constant in time. Although in practical situations there are significant variations with respect to these nominal conditions, the controller is expected to tackle this difficulty.

Figures 7.6, 7.7, 7.8 and 7.9 show the results obtained with the servo controller described. This simulation is performed with a detailed model of the DCSF that incorporates dynamics and nonlinearities not present in Eq. (7.1). Furthermore, disturbances are injected in order to test the servo controller in nonideal conditions.

Figure 7.6 shows the temperatures along the pipe, and their respective references generated by the motion planner. The references are constant and increase with the distance from the pipe inlet. After an initial transient in which the plant temperatures converge to their references, the tracking errors become small.

Figure 7.7 shows the nominal radiation used for planning and the radiation actually measured. Their differences are a source of disturbances that are compensated by the feedback controller. Furthermore, at about $t = 2.5$ h a disturbance of 20 °C was injected during 150 s in the inlet fluid temperature, thereby creating another source of disturbance. Figures 7.8 and 7.9 show, respectively, the plant outlet fluid temperature and reference to track and the corresponding planned fluid flow and fluid flow yielded by the closed loop.

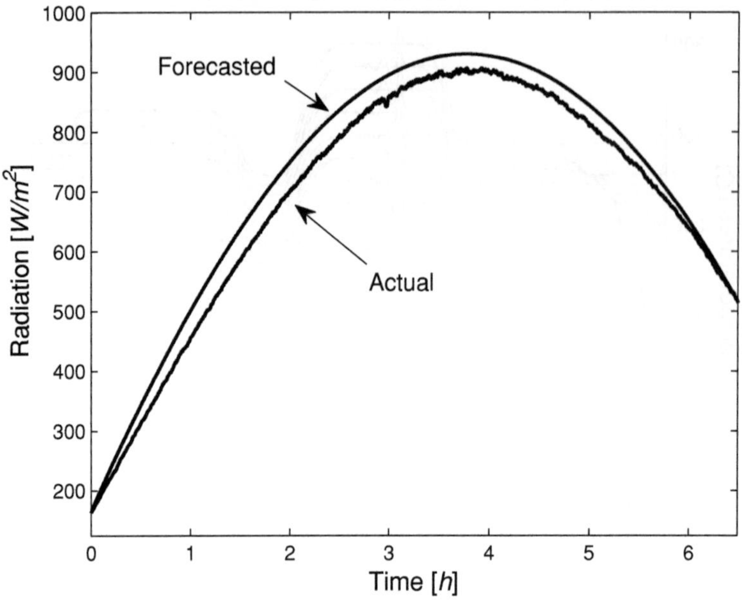

Fig. 7.7 Nominal radiation used for planning (*upper, smoother curve*) and radiation actually measured (*lower, noisy curve*)

The planned fluid, shown in Fig. 7.9 is the thick line. The fluid flow that is actually applied to the DCSF normally tracks closely the planned flow, with deviations accounting to compensate the disturbances. For instance, the correction introduced by the feedback controller is apparent after time 2, 5 h to compensate for the disturbances injected in the inlet fluid temperature.

7.3 Adaptation

The parameter α in Eq. 7.1 reflects not only the mirror efficiency but is also influenced by the nonlinear dependence of fluid-specific heat on temperature. Changing the operating point (as in the servo problem) causes thus changes of this parameter. With this motivation, the servo control law was modified in order to incorporate the estimate of α. Accordingly, the control law is now given by

$$u(t) = \frac{\hat{\alpha}(t)R(t)}{\tilde{f}(t)} - \frac{K_C}{\tilde{f}(t)}e(t), \tag{7.55}$$

where $\tilde{f}$ is the planned input, assuming that $\alpha R_{\text{nom}}(t) = 1$. Furthermore, the estimate $\hat{\alpha}$ of α is updated according to

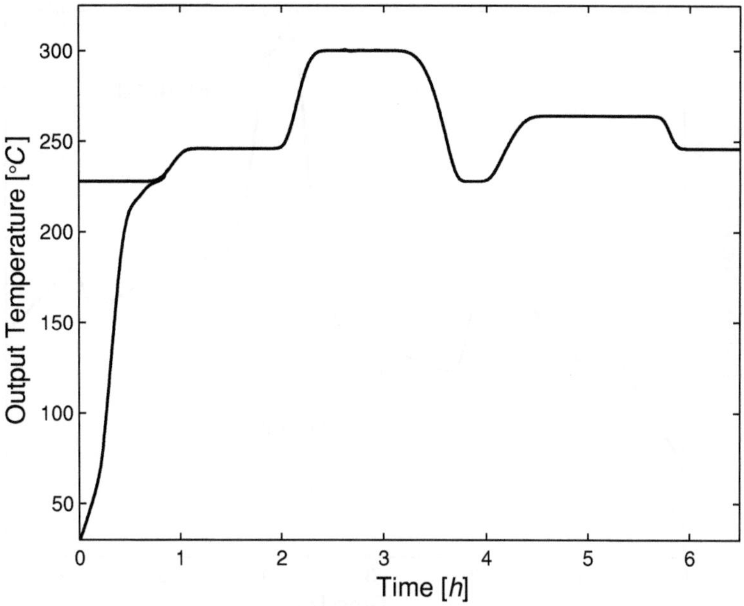

Fig. 7.8 Closed loop: Plant outlet fluid temperature and the reference to track

$$\hat{\alpha}(t) = -\gamma \int_0^t R(\zeta)e(\zeta)d\zeta + \hat{\alpha}(0). \tag{7.56}$$

In this way a PI control law is obtained, incorporating at the same time an estimate of α.

For justifying the above-adaptive control law, an argument based on the approximation of (7.1) by a lumped parameter model is presented. For that sake, consider N points along the pipe, equally distant in space at positions jh, $h = L/N$ $j = 1, \ldots, N$ and let T_j be the temperature at the corresponding point. Then, as discussed in Sect. 2.4.1 , Eq. (7.1) can be approximated by a set of ODE's of which the last one is

$$\dot{T}_N = -\frac{u(t)}{h}(T_N - T_{N-1}) + (\hat{\alpha} - \tilde{\alpha})R, \tag{7.57}$$

where

$$\tilde{\alpha} = \hat{\alpha} - \alpha. \tag{7.58}$$

Inserting the value of u given by (7.55) on (7.57) yields the closed-loop model

$$\dot{T}_N = -\left[\frac{\hat{\alpha}(t)R(t)}{\tilde{f}(t)} - \frac{K_C}{\tilde{f}(t)}e(t)\right]\frac{T_N - T_{N-1}}{h} + (\hat{\alpha} - \tilde{\alpha})R. \tag{7.59}$$

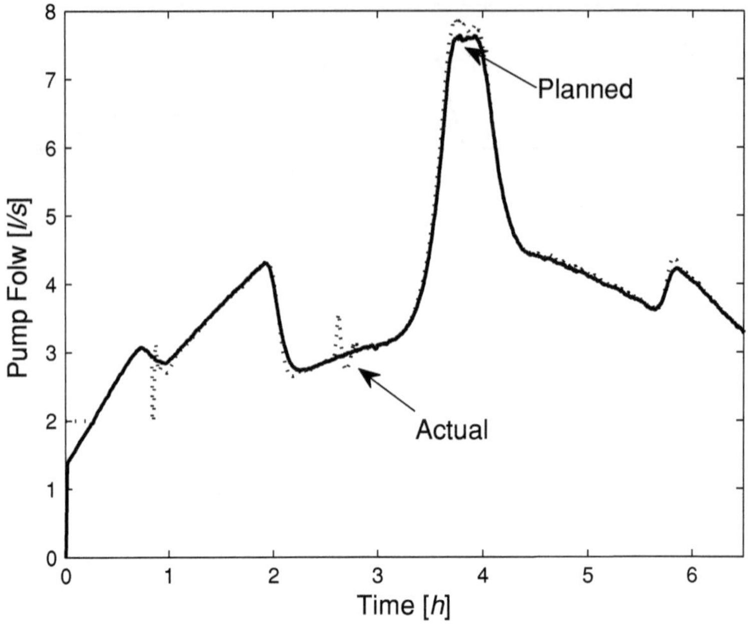

Fig. 7.9 Planned fluid flow (the *thick line*) and fluid flow yielded by the closed loop [m^3/s]

Since the planned input is such that

$$\frac{T_N - T_{N-1}}{h} \approx \tilde{f}, \tag{7.60}$$

it finally comes that the outlet fluid temperature verifies the ODE

$$\dot{T}_N = K_C e(t) + \tilde{\alpha} R. \tag{7.61}$$

From this equation, by repeating the arguments in Chap. 6 based on a joint Lyapunov function for estimation and control, it is then possible to obtain the adaptation rule, while proving that the overall system is stable and that the tracking error will tend to zero.

Figure 7.10 shows the detailed block diagram of the adaptive motion planner just described, and Figs. 7.11, 7.12 and 7.13 illustrate its application to a DCSF.

Figure 7.11 shows the results obtained for the manipulated variable, while Fig. 7.12 shows the outlet fluid temperature and the reference to track. After an initial adaptation transient, the outlet fluid temperature tightly tracks the reference.

Since the computation of the fluid flow u from the virtual manipulated variable f includes a dependency on the solar radiation (as seen from (7.10)), a feedforward effect that compensates radiation changes is included. The existence of feedforward means that, when there is a drop in incident radiation, the fluid is immediately

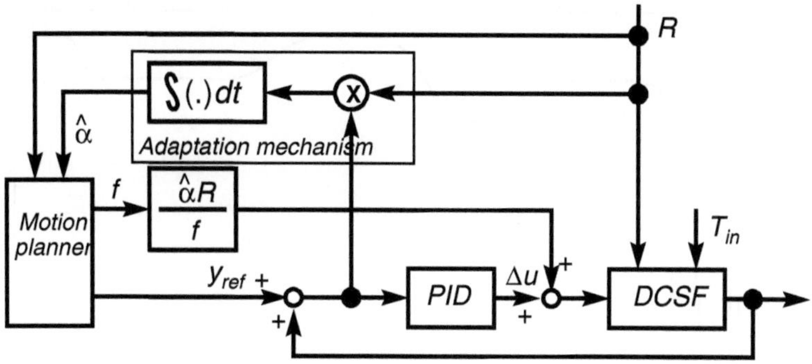

Fig. 7.10 Detailed structure of the adaptive motion planner

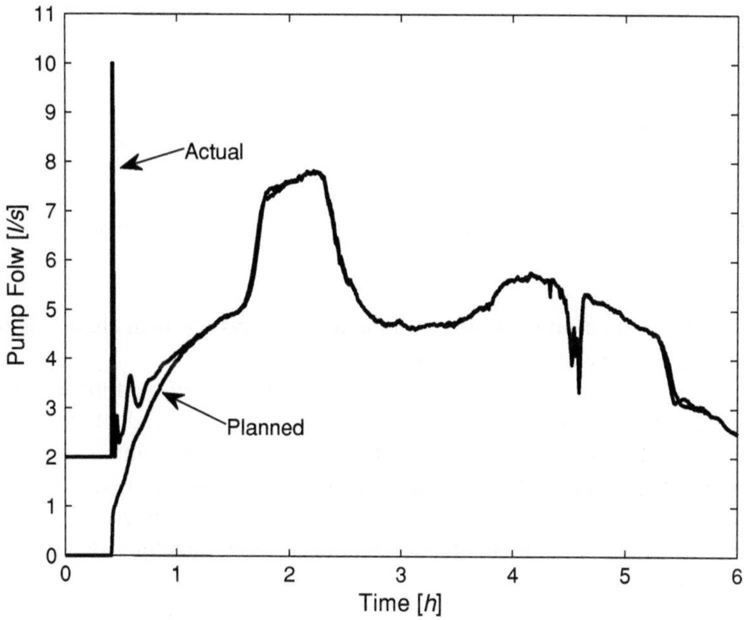

Fig. 7.11 Adaptive servo. Planned fluid flow and the fluid flow yielded by the closed loop with adaptation [m^3/s]

reduced, even before there is a drop in the outlet fluid temperature. This effect is noticeable in Fig. 7.11 at about time 4, 5 h, where the fluid is reduced to compensate a sudden drop on solar radiation caused by a passing cloud.

Of course, this feedforward effect depends on a reliable measurement of the solar radiation that attains the DCSF. It may happen that the radiation sensor is located away from the field (e.g., if it serves as an instrument that provides support to different plants in a complex), or that the DCSF is so large that different parts of it might be illuminated differently. In this situation, the feedback controller is expected to tackle

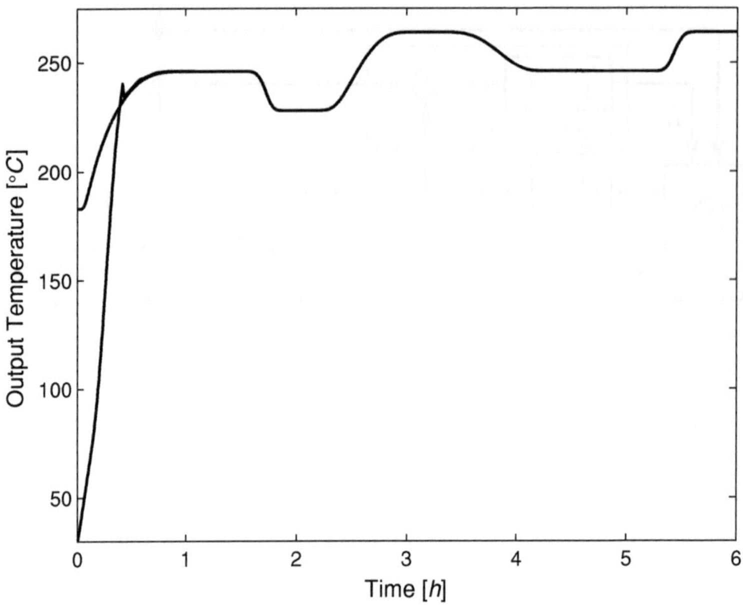

Fig. 7.12 Adaptive servo: Plant outlet fluid temperature and the reference to track

the situation, compensating the errors of the feedforward action in order to still keep the outlet fluid temperature close to the reference.

Figure 7.13 (above) shows the convergence of the parameter α estimate. The initial estimate is away from its correct value, a fact that originates an adaptation transient, during which the estimate gradually approaches its correct value. Although this adaptation process causes a transient on the manipulated variable, as seen in Fig. 7.11, no unacceptable transients are observed in the outlet fluid temperature, as seen in Fig. 7.12. Of course, initializing the estimate of α close to its final value avoids the adaptation transient.

After the estimate of α approaches its correct value, it is noticeable in Fig. 7.13 (above) that its value is affected by changes in the reference. Nevertheless, these variations have a negligible influence on the output of the controlled system.

Figure 7.13 (below) plots the quotient

$$\frac{T_N - T_{N-1}}{h\tilde{f}}.$$

After the initial adaptation transient, it becomes close to 1, thereby verifying the assumption underlying (7.60).

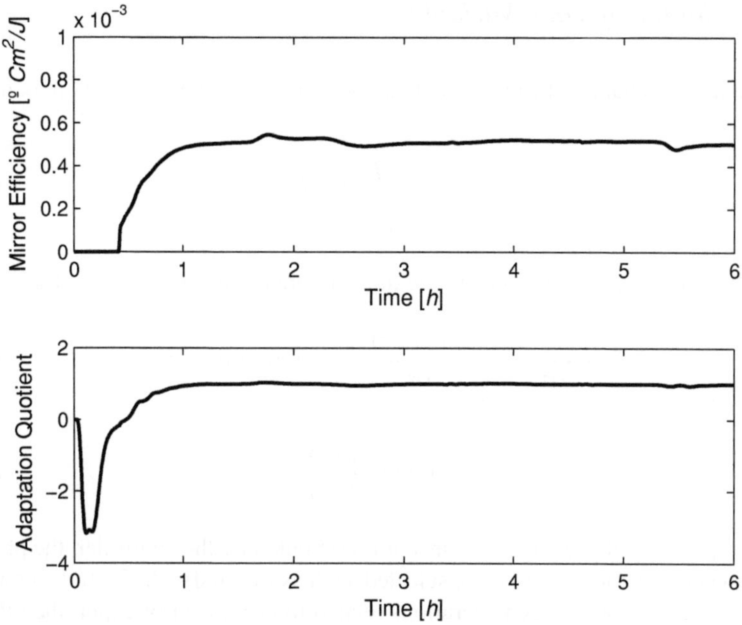

Fig. 7.13 Adaptive servo. *Above*: Parameter α (mirror efficiency) estimate convergence ($\hat{\alpha}(0) = 0$). *Below*: Checking the condition for stability analysis

7.4 Motion Planning of a Moisture Control System

Consider now the motion planning problem for the moisture control system described in Sect. 2.5.1. Remember that this process is described by the PDE model

$$\frac{\partial w(z,t)}{\partial t} + v(t)\frac{\partial w(z,t)}{\partial z} = -\beta\, q(t)\, w(z,t) \tag{7.62}$$

where $z \in [0, L]$ is the length measured along the conveyor that carries the product to be dried, $t \in [\,0, +\infty\,[$ is the continuous time, $w(z, t)$ is the moisture content of the product per unit volume, $v(t)$ is the velocity of the conveyor belt that carries the product, $q(t)$ is the flow of hot air and β is a parameter.

The motion planning problem corresponds to find the manipulated variable $v(t)$ that drives the space function that describes the moisture content along the conveyor between two given values corresponding to stationary states (i.e., a distribution of moisture along the belt that keeps constant in time for constant values of the velocity and hot air flow). The solution to this problem is obtained as in the motion planning problem for a DCSF, by making a change of time variable using the concept of orbital flatness, and finding, in the new coordinate space $[z, \tau]$, a flat output.

7.4.1 Change of Time Variable

By making the change of time variable associated to the conveyor belt velocity

$$\tau(t) = \int_0^t v(\sigma)d\sigma, \tag{7.63}$$

the PDE (7.62) that corresponds to the moisture model is written in the form

$$\frac{\partial w(z,\tau)}{\partial \tau} + \frac{\partial y(w,\tau)}{\partial z} = -\beta\, u(\tau)\, w(z,\tau), \tag{7.64}$$

where

$$u(\tau) \triangleq \frac{q(\tau)}{v(\tau)} \tag{7.65}$$

is a virtual manipulated variable. Once u is computed by the controller, the physical manipulated variables q and v are selected such as to satisfy (7.65). If the conveyor velocity v is imposed by considerations related to process throughput, then the hot air flow q is given by:

$$q(\tau) = v(\tau)u(\tau). \tag{7.66}$$

7.4.2 The Homogeneous Equation

The homogeneous equation in transformed coordinates is obtained by making $u(\tau) = 0$ in (7.64), and is given by:

$$\frac{\partial y(z,\tau)}{\partial \tau} + \frac{\partial y(z,\tau)}{\partial z} = 0. \tag{7.67}$$

As in the case of the DCSF model, let $\phi(\xi)$ be an arbitrary function, C a constant, and

$$\xi = \tau - z + C. \tag{7.68}$$

Since the direct replacement in the homogeneous Eq. (7.67) results in a true equality, it is concluded that its general solution is given by:

$$\phi(\tau - z + C). \tag{7.69}$$

7.4.3 Solution of the Nonhomogeneous Equation

From the solution of the homogeneous equation, it is readily proved that the solution of the nonhomogeneous equation in transformed coordinates is given by:

$$\psi(x,\tau) = \phi(x)e^{-\beta U(\tau)}, \tag{7.70}$$

where $U(\tau)$ is a primitive function of $u(\tau)$. Indeed, observing that

$$\frac{\partial\psi(x,\tau)}{\partial\tau} = \frac{\mathrm{d}\phi}{\mathrm{d}x}\, e^{-\beta U(\tau)} + \phi(x)(-\beta u(\tau))\, e^{-\beta U(\tau)} \tag{7.71}$$

and

$$\frac{\partial\psi(x,\tau)}{\partial z} = -\frac{\mathrm{d}\phi}{\mathrm{d}x}\, e^{-\beta U(\tau)}, \tag{7.72}$$

and inserting these derivatives in (7.64), a true equality is obtained. Hence, from (7.68) and (7.70), it is concluded that the general solution of the nonhomogeneous Eq. (7.64) is written as:

$$w(z,\tau) = \phi(\tau - z + C)\, e^{-\beta U(\tau)}, \tag{7.73}$$

where C is a constant that depends on the boundary conditions.

7.4.4 Flat Output

For the moisture control process, define the flat output $\bar{y}$ by the following boundary condition, imposed for $z = L$ in the transformed time coordinate:

$$\bar{y}(\tau) \stackrel{\Delta}{=} \left.\frac{\partial \ln\psi(x,\tau)}{\partial z}\right|_{z=L}. \tag{7.74}$$

There is not an algorithm to obtain a flat output. Instead, one has to be lead by intuition, in a trial-and-error procedure, and verify a posteriori wether a given function is actually a flat output. In this case, we get inspiration from a comparison with the flat output obtained for the DCSF that, as shown before, is the spacial gradient of the temperature at the pipe outlet (i.e., for $z = L$. Therefore, we seek as a flat output a gradient of a function of the moisture for $z = L$ and, given the exponential decay in the expression of the general solution (7.73), we take this function as a logarithm. This reasoning explains how the choice (7.74) pops out. In the sequel, we prove that $\bar{y}(\tau)$ defined by (7.74) is actually a flat output. This proof is done by showing that the manipulated input that drives the moisture content $w(z, L)$ along a specified trajectory can be computed as a linear combination of the assumed flat output and its derivatives.

To proceed, start by observing that, using (7.70) and the definition of the flat output (7.74), it follows that

$$\bar{y}(\tau) = \frac{\partial \ln \psi(x,\tau)}{\partial x}\frac{\partial x}{\partial z}\bigg|_{z=L}, \tag{7.75}$$

or

$$\bar{y}(\tau) = \frac{\partial}{\partial x}(\ln\phi(x) - \beta U(\tau))\frac{\partial x}{\partial z}\bigg|_{z=L} = \frac{\mathrm{d}\ln\phi(x)}{\mathrm{d}x}(-1)\bigg|_{z=L}, \tag{7.76}$$

and it is finally concluded that

$$-\bar{y}(\tau) = \frac{1}{\phi(x)}\frac{\mathrm{d}\phi(x)}{\mathrm{d}x}\bigg|_{z=L} = \frac{1}{\phi(\tau)}\frac{\mathrm{d}\phi(\tau)}{\mathrm{d}\tau}, \tag{7.77}$$

where $z = C = L$.

The solution of the differential equation (7.77) yields the particular function $\phi(\tau)$ that corresponds to the flat output defined in (7.74) as:

$$\phi(\tau) = e^{-\bar{W}(\tau)}, \tag{7.78}$$

where $\bar{W}(\tau)$ is a primitive of $\bar{w}(\tau)$. The solution (7.73) of the nonhomogeneous equation (7.64) with a ϕ that corresponds to the flat output (7.74) is thus

$$\bar{w}(z,\tau) = e^{-\bar{W}(\tau-z+L)}e^{-\beta U(\tau)}. \tag{7.79}$$

We write the bar in $\bar{w}$ to emphasize the fact that (7.79) is a solution of the general Eq. (7.64) that verifies (7.74). We call $\bar{w}$ a flat solution.

7.4.5 Planning Formulas

Consider now the boundary condition for $z = 0$. By Making $z = 0$ in (7.79) and taking the logarithm, we get

$$\ln\bar{w}(0,\tau) = -\bar{W}(\tau+L) - \beta U(\tau). \tag{7.80}$$

Combining (7.80) and (7.79) yields

$$\ln\left(\frac{\bar{w}(z,\tau)}{\bar{w}(0,\tau)}\right) = \bar{W}(\tau+L) - \bar{W}(\tau-z+L). \tag{7.81}$$

Furthermore, for $w(0,\tau) = w_0$ constant, differentiation of (7.80) with respect to τ yields

$$u(\tau) = -\frac{1}{\beta}\,\bar{w}(\tau + L). \tag{7.82}$$

Finally, the development of $\bar{y}(\tau+L)$, $\bar{W}(\tau+L)$ and $\bar{W}(\tau-z+L)$ in a Taylor series yields

$$\ln\left(\frac{\bar{w}(z,\tau)}{w_0}\right) = \sum_{k=1}^{\infty}\frac{L^k}{k!}\overset{(k-1)}{\bar{y}}(\tau) - \sum_{k=1}^{\infty}\frac{(L-z)^{k-1}}{k!}\overset{(k-1)}{\bar{y}}(\tau) \tag{7.83}$$

and

$$u(\tau) = -\frac{1}{\beta}\sum_{k=0}^{\infty}\frac{L^k}{k!}\overset{(k-1)}{\bar{y}}(\tau) \tag{7.84}$$

where $\overset{(i)}{\bar{w}}(\tau)$ is the derivative of order i of the flat output. The formulas (7.83) and (7.84) express the process output and input as a function of the flat output derivative and of its derivatives, provided that the series converge.

Motion planning is made by making moves between successive stationary states defined by

$$\frac{\partial w_{ss}(z,t)}{\partial t} = \frac{\partial w_{ss}(z,\tau)}{\partial \tau} = 0, \tag{7.85}$$

and a constant value u_{ss} for the manipulated input. This condition implies that

$$\frac{\mathrm{d}w}{\mathrm{d}z} = -\beta\, u_{ss}\, w(z), \tag{7.86}$$

or

$$w_{ss}(z) = w_{ss}(0)\, e^{-\beta u_{ss} z}, \tag{7.87}$$

where $w_{ss}(0)$ is the boundary condition at $z = 0$, and $u_{ss} = \frac{q_{ss}}{v_{ss}}$, are constant values ($v_{ss}$is the constant velocity of the conveyor belt and q_{ss} is the constant flow of hot air that define the equilibrium). By making

$$w_{ss}(0) = w_0, \tag{7.88}$$

a constant, we get for two consecutive stationary states

$$w_1(L) = w_1 = w_0\, e^{-\beta u_1 L} \tag{7.89}$$

$$w_2(L) = w_2 = w_0\, e^{-\beta u_2 L} \tag{7.90}$$

and

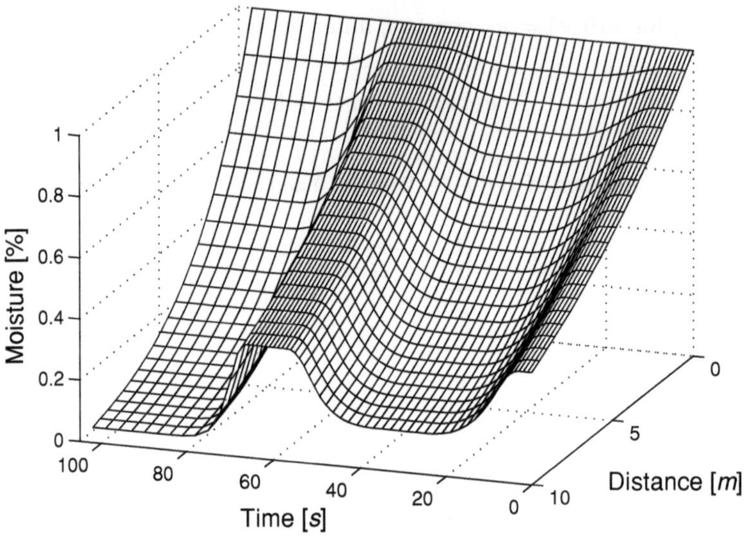

Fig. 7.14 Planning the moisture dryer output with multiple transitions

$$u_1 = -\frac{1}{\beta L}\ln\left(\frac{y_1}{y_0}\right) = -\frac{1}{\beta}C_1 \tag{7.91}$$

$$u_2 = -\frac{1}{\beta L}\ln\left(\frac{y_2}{y_0}\right) = -\frac{1}{\beta}C_2. \tag{7.92}$$

These formulas allow to establish a plan for moving between the stationary states using the flat output. In order to allow a smooth transition, an exponential Gevrey function of class α, denoted $\Phi_{y\sigma}$, is used, as in the DCSF case study, being then

$$\bar{y}(\tau) = C_1 + (C_2 - C_1)\Phi_{y\sigma}\left(\frac{\tau}{\tau^*}\right) \tag{7.93}$$

$$\Phi_{y\sigma}(0) = 0 \quad \tau \leq 0 \tag{7.94}$$

$$\Phi_{y\sigma}(1) = 1 \quad \tau \geq \tau^* \tag{7.95}$$

An example of using this expression to move the state of the moisture control system among different stationary states is given in Fig. 7.14. In these cases, the series (7.84) was approximated by its first 20 terms.

7.5 Main Points of the Chapter

This chapter shows how to develop an adaptive servo controller for a DCSF, using a motion planner based on orbital flatness. The structure of the controller considered in this chapter is shown in Fig. 7.1 and comprises three main blocks:

- A motion planner;
- A feedback controller;
- An adaptation law.

The motion planner acts like a feedforward controller and computes in open loop a function that defines the manipulated variable such as to drive the DCSF output (given by the outlet fluid temperature) between a specified sequence of stationary states (that can be considered as way points if a parallel with Robotics is made). The transitions between two consecutive stationary states follow a trajectory defined by a particular class of functions, known as exponential Gevrey functions of class α. These functions allow to perform smooth transients between stationary states of a DCSF, while avoiding some difficulties inherent to the use of analytical functions, and at the same time preserving the convergence of some series that are required for the solution of the motion planning problem.

The feedback controller modifies the value of the manipulated variable yielded by the motion planner in order to compensate the disturbances that act on the plant, including the effects of unmodeled dynamics, and keeping the process output close to its reference, also generated by the motion planner. This controller amounts to be either a linear state feedback with a variable sampling time, or a PID.

Finally, the adaptation law adjusts the estimate of a parameter that depends both on collector mirror efficiency and on the thermodynamic properties of the fluid, and thereby changes with the operating point and due to unpredictable factors (as discussed in Sect. 1.3. As in Chap. 6, the estimate of this parameter is obtained with a control Lyapunov function.

The central issue in the present chapter is the use of a flat output to perform motion planning. A flat output is a variable such that all the variables that characterize the system (states and manipulated variables) can be computed as an algebraic function of the flat output and its derivatives. Although for a DCSF there is no flat output available, it is possible to perform a change in the time scale, and to redefine the manipulated variable, such that the transformed system has a flat output that, for the DCSF is the gradient of the outlet fluid temperature. This is the concept of orbital flatness.

The change of time scale considered is associated to the fluid flow, and renders the DCSF model linear with respect to a transformed manipulated variable. This change of time scale plays a major role in the present chapter, as well as in Chap. 5.

In addition to the solution of the adaptive servo problem for an infinite-dimensional model of a DCSF, similar techniques have been applied to solve the motion planning problem of a moisture control system.

7.6 Bibliographic Notes

The concept of flat systems has been introduced (Fliess et al. 1995) to generalize the concept of controllable linear systems to a subset of nonlinear systems. Concepts associated with flatness allow to solve the problem of inverting the plant model in order to find a control function that steers the state from one point to another such as to perform a desired task (Martin et al. 1997; Fliess et al. 1995, 1999). Flatness is thus associated to the solution of nonlinear control problems in areas that range from robotics and vehicle motion planning to chemical reactors (Respondek et al. 2004; Rothfuss et al. 2014; Rudolph and Mounier 2000).

The trajectory specified that links the initial value of the output with the final value is to be smooth. If all the trajectory derivatives are zero at the beginning and at the end of the interval during which the transition takes place, the function to be chosen cannot be analytic because, otherwise, the function itself would be zero. On the other way, the function used must be such that the series that define the control converge. Gevrey functions of class α, with $\alpha > 1$ are used for this propose (Lynch and Rudolph 2002). In Rudolph et al. (2003) several computational aspects of Gevrey functions are addressed from a practical point of view.

Although originally designed for finite-dimensional nonlinear systems, the approach has been extended to infinite-dimensional systems as well (Lynch and Rudolph 2002; Mounier and Rudolph 1998; Rudolph 2003; Rudolph et al. 2003; Rouchon 2001). Examples include delay systems (Mounier and Rudolph 1998) and systems described by parabolic partial differential equations (Lynch and Rudolph 2002). The monographs (Rudolph 2003; Rudolph et al. 2003) provide a comprehensive account of flatness-based control of distributed systems, with many examples from various fields with technological interest.

Some infinite-dimensional systems, in particular of hyperbolic type, become flat with a change of time variable, such as considered in this book in Chaps. 5 and 7. This is the concept of orbital flatness (Guay 1999; Respondek 1998). An application of the concept of orbital flatness to the control of batch cooling crystallization processes is described in Vollmer and Raisch (2003). The solution of the adaptive motion planning for the DCSF plant using orbital flatness was originally presented in Igreja et al. (2004).

The concept of orbital flatness is related to the concept of *substantial derivative*, known in Physics. The substantial derivative, also called *material derivative* (Bird et al. 2007) is the rate of change of a physical quantity as measured by an observer that is moving along the flow of a fluid element.

References

Bird RB, Stewart WE, Lightfoot EN (2007) em Transport phenomena, 2nd edn. Wiley, Weinheim

Fliess M, Lévine J, Martin P, Rouchon P (1995) Flatness and defect of nonlinear systems: introductory theory and examples. Int J Control 61(6):1327–1361

Fliess M, Lévine J, Martin P (1999) A Lie-Bäcklund approach to equivalence and flatness of nonlinear systems. IEEE Trans Autom Control 44:922–937

Fliess M, Lévine J, Martin P, Rouchon P (1995) Design of trajectory stabilizing feedback for driftless flat systems. In: Proceedings of the 3rd ECC, Rome, Italy, pp 1882–1887

Guay M (1999) An algorithm for orbital feedback linearization of single-input control affine systems. Syst Control Lett 38:271–281

Henson M, Seborg D (1997) Nonlinear process control. Prentice Hall, New Jersey

Igreja JM, Lemos JM, Rouchon P, Silva RN (2004) Dynamic motion planning of a distributed collector solar field. In: NOLCOS 2004, Stuttgart, Germany

Lewis FL, Syrmos VL (1995) Optimal control. Wiley, New York

Lynch AF, Rudolph J (2002) Flatness-based boundary control of a class of quasilinear parabolic distributed parameter systems. Int J Control 75(15):1219–1230

Martin Ph, Murray RM, Rouchon P (1997) Flat systems. In: Bastin G, Gevers M (eds) Plenary lectures and mini-courses—European control conference

Mosca E (1995) Optimal, predictive, and adaptive control. Prentice Hall, New Jersey

Mounier H, Rudolph J (1998) Flatness-based control of nonlinear delay systems: a chemical reactor example. Int J Control 71(5):871–890

Respondek W (1998) Orbital feedback linearization of single-input nonlinear control systems. In: Proceedings of the NOLCOS'98, Enschede, The Netherlands, pp 499–504

Respondek W, Pogromsky A, Nijmeier H (2004) Time scaling for observer design with linearizable error dynamics. Automatica 40:277–285

Rothfuss R, Rudolph J, Zeita M (2014) Flatness based control of a nonlinear chemical reactor model. Automatica 32:1433–1439

Rouchon P (2001) Motion planning, equivalence, infinite dimensional systems. Int J Appl Math Comput Sci 11:165–188

Rudolph J (2003) Flatness based control of distributed parameter systems. Shaker Verlag, Aachen

Rudolph J, Mounier H (2000) Trajectory tracking for pi-flat nonlinear delay systems with a motor example. In: Isidori A et al (eds.) Nonlinear control vol 2. Springer, Berlin, pp 339–351

Rudolph J, Winkler J, Woittennek F (2003) Flatness based control of distributed parameter systems: examples and computer exercises from various technological domains. Shaker Verlag, Aachen

Vollmer U, Raisch J (2003) Control of batch cooling crystallization processes based on orbital flatness. Int J Control 76(16):1635–1643

Chapter 8
Conclusions

What are the essential conclusions to be drawn from this book? Hereafter, we start by drawing a "road map" for the DCSF adaptive control algorithms described in the previous chapters. This road map explains what to expect from the algorithms in what concerns their achievable performance as well as their capacity to be used with more or less modifications in different plants that share with DCSF's the fact that they rely essentially on transport phenomena along one space dimension. Examples are provided by steam super-heaters in thermoelectric power plants, moisture control systems, tubular chemical reactors, water delivery canals, and motorway traffic. The models of DCSFs discussed in Chap. 2 are instrumental in this respect, specially in what concerns the control algorithms of Chaps. 5 and 7 (that perform a model linearization resorting to a change of the time scale), and 6 (that rely on feedback linearization and control Lyapunov functions).

After understanding this picture, the utility of the algorithms is discussed in the framework of energy production incorporating renewable sources. Exploring this field of work implies being able to tackle difficulties like optimization in the presence of high levels of uncertainty, time-varying dynamics and planning over extended periods of time, for which the control algorithms described in Chaps. 3–7 provide valuable contributions.

The final section establishes the "boundary conditions" of the book by recalling a list of topics that are outside the scope of the matters being treated, but which they are important either when pursuing applications with a broader view, in particular in the field of renewable energy production, or when extending the algorithms and the study of their properties.

8.1 Road Map of DCSF Adaptive Control Algorithms

As discussed in Chap. 1, solar energy systems are quite wide in scope and they are currently the subject of a renewed interest associated to the search for renewable sources of energy that may form an alternative, or a complement, to other types

J. M. Lemos et al., *Adaptive Control of Solar Energy Collector Systems*, Advances in Industrial Control, DOI: 10.1007/978-3-319-06853-4_8,

of energy sources, such as fossil fuels or nuclear, or even other renewable energy sources.

Motivated by this interest, this book addresses the control of distributed collector solar fields (DCSFs). This technology allows to implement solar thermal plants that produce a significant amount of electric power, of the order of tens or even hundreds of megawatts, and is specially attractive for places such as deserts, where the cost of space is very low and the number of solar hours per year is very high. As such a number of commercial plants have already been build in the Middle East, southern Spain, or close to Las Vegas, in the USA.

8.1.1 DCSF Characteristics and Control Design

Due to their large dimensions and specificity of operation, where a fluid circulates through a pipe located at the focus of solar energy collector mirrors, the dynamics of DCSFs depends on space as well as on time. Together with uncertainty sources in plant model, this dependency raises interesting challenges for controlling the temperature on DCSFs and motivates the consideration of the adaptive control methods addressed in this book.

The first step in designing a control system is to identify what are the plant outputs that relates to the control objectives, the manipulated variable (the variable that can be directly changed and that allows to exert an influence on the output variable), and to characterize the disturbances. Identifying the models that represent the dynamic and static relationship among these variables is also essential to improve performance. It is also advantageous to discriminate between unmeasurable disturbances (unknown input signals that cause a deviation of the process output from the desired reference) and accessible disturbances, that are disturbances that can be measured and used as an input to improve controller performance. Chapter 1, and in particular Sect. 1.3, addresses the above issues and provides a qualitative insight on the control system structure associated to a DCSF. In broad terms, DCSF controllers combine feedback from the fluid temperature to settle it at the desired reference value with a feedforward action from the solar radiation that constitutes the main accessible disturbance.

Chapter 2 proceeds then to deduce the mathematical models that relate the above variables in a DCSF. These models are of different type. If the model is to allow the computation of the temperature at all the points along the pipe, the model takes the form of a DCSF. If, instead, one considers only some points along the pipe, a set of nonlinear ODEs results. Furthermore, Chap. 2 addresses the type of dynamic response to expect from the temperature when the manipulated variable is changed. Both these models and the insight they provide on the DCSF dynamic response are used in subsequent chapters that are devoted to adaptive controllers.

In this book, instead of estimating the model parameters from data collected *a priori* from the plant, and then designing the controller from the resulting models, model identification and controller design are performed online, in an interlaced way. Therefore, when a data sample is read at a given sampling instant, it is used both to update the model parameters and to compute the manipulated variable using controllers which gains are redesigned according to the new parameter estimates. This so-called "Adaptive Control" approach has a number of advantages, but also drawbacks.

In general, two main advantages of Adaptive Control are

- The fast prototyping and commissioning of the controller;
- The capacity of the controller to track plant model parameter changes, that may happen due to a variety of reasons such as equipment wearing due to aging or change of operational conditions. This capacity yields an increase along time of control performance.

Although these advantages are common to a wide class of plants, it is discussed in Sect. 1.3 that, in the special case of DCSFs, its dynamics may vary in unpredictable ways due to a number of factors that are inherent to this type of plants, such as dust deposition on mirrors or mirror deformation due to wind. The use of adaptive control is thus quite natural because it allows to redesign the controller online, so as keep it matched to the DCSF dynamics.

Three main drawbacks of adaptive control are

- Lack of a complete theory that ensures stability on realistic grounds;
- Possibility of adaptation transients that are unacceptable from an operational point of view;
- Possibility of the adaptation mechanism to drive the controller gains to unsafe operating regions, or even to a situation in which the controlled system might become unstable.

8.1.2 Controllers Based on Data-Driven Models

The type of adaptive controller used may vary a lot. In this book, a major trade-off is concerned with the amount of plant knowledge embedded on the algorithm. At one extreme there are algorithms, addressed in Chap. 3, which assume that the plant is linear. Their main drawback consists in not taking into account plant nonlinearities when performing maneuvres that consist of large and relatively fast excursions of the output, or when rejecting fast disturbances, which is more common on DCSF operation. In DCSFs, these algorithms are mainly useful for temperature regulation around piecewise constant operating points. They can also be useful parts of the multiple model-based algorithms described in Chap. 4.

The most powerful algorithms in this class are obtained by combining a linear model predictive controller (MPC) with an adaptation mechanism using the so-called certainty equivalence principle. Depending on the assumptions made, different algorithms are yielded. Since they compute the manipulated variable on the basis of an optimization of a quadratic cost defined over an extended horizon, they inherit the advantages (and drawbacks as well) of linear quadratic control. In particular, if the time horizon over which the cost is defined is large enough, a stabilizing controller for the nominal model is yielded. A well-known algorithm of this class is the adaptive generalized predictive controller, or GPC in short, described in Sect. 3.4. When making the assumption that the plant is being controlled under a fixed gain controller, a simplified linear model results, yielding the MUSMAR controller, discussed in Sect. 3.6. Compared with GPC, MUSMAR has the advantages of minimizing locally the control cost progressing along a Newton direction and of requiring a shorter horizon. However, MUSMAR has the drawback of using predictors that implicitly assume that a state feedback with a constant gain is applied to the plant. This last assumption is violated, e.g., when the manipulated variable saturates frequently, causing MUSMAR to diverge if proper actions (that amount to blocking the adaptation mechanism) are not taken under these situations.

As explained in Chap. 3, adaptive MPC algorithms based on linear predictive models, with adaptation relying on the certainty equivalence principle are useful for fast controller prototyping and optimization since they require very little prior knowledge from the plant. As discussed, they can be useful with other types of plants as well. Among other case studies, Sect. 3.7.2 presents an example in which MUSMAR is able to minimize superheated steam temperature fluctuations in a way that is directly connected to economic process performance.

Another class of algorithms that addresses the adaptive control problem for a wide class of plants is supervised multiple model adaptive control (MMAC), the essentials of which, when applied to DCSS, are presented in Chap. 4. The controller structure consists of a bank of candidate local controllers that are designed to match a corresponding bank of local linear models that "cover" the possible outcomes of plant dynamics. In MMAC, the output of the candidate local models is compared online with observed plant output data in order to decide which local model best matches the plant dynamics at a given time. The local controller that is actually applied to the plant is the one that corresponds to the best match, but ensuring a so-called "dwell time" condition. Indeed, since it is possible to obtain an unstable time-varying dynamic system by switching among stable dynamic systems, when a controller is applied to the DCSF, it remains so for at least a minimum dwell time, a condition that ensures stability of the controlled system.

The choice of the local models is an important issue in MMAC. In DCSF control, one may be guided by the physical intuition about the dependence of plant dynamics on variables such as the range of output temperature, heating fluid flow, and solar radiation intensity to decide which models are needed and what are the operational conditions needed to identify them. Although general methods to identify hybrid systems are available (Paoletti 2007), one should bear in mind that the ultimate goal is not to replicate the plant output with a model, but to design a controller that stabilizes the overall system and optimizes performance.

Too few local models will lead to a degradation of performance since the corresponding local controllers must be conservative in order to stabilize the plant over enlarged operating regions. There are examples in which removing a pair model/controller that is applied only for short periods of time leads to a noticeable performance degradation. On the other side, a big increase on the number of local controllers may lead to excessive switching that induces fast variations on the manipulated variable. As mentioned in Chap. 4, metrics that compare dynamic models bearing a relationship with stability, such as discussed in Anderson et al. (2000) provide an adequate formal means to select local controllers.

8.1.3 Exploiting DCSF Structure

In MMAC, at least implicitly, local linear controllers are patched together to form an approximation to a global nonlinear controller. It is thus possible to take advantage of this fact to compensate some plant nonlinearities. This compensation is only approximate because it relies on a finite number of local controllers being patched. Furthermore, it may not be clear how to design the local controllers in a systematic way to achieve this objective. The examples provided for both a pilot air heating turbofan in Sect. 4.3 and for a DCSF in Sect. 4.4 that rely on the variables that define the operational regimes of these plants, and qualitative physical insight, provide clues for designing this type of controllers.

A much better compensation of the nonlinearities requires the DCSF models discussed in Chap. 2. One possible approach along this line consists of making a change of time variable that depends on the fluid flow. In practice, this transformation can be implemented in a computer control framework with a sampling interval that varies with the flow. Together with the change in the time variable, a virtual manipulated variable is computed such that the plant model becomes linear. The controller that computes this virtual manipulated variable is thus simplified. In turn, the physical manipulated variable (fluid flow) can be easily computed from this virtual control. This approach is followed in Chap. 5 to obtain nonlinear MPC controllers, named WARTIC-i/o and WARTIC-state, that allow to make fast jumps of the outlet fluid temperature almost with no overshoot. Adaptation is embedded according to a certainty equivalence assumption, with the control laws being computed with models that depend on parameter estimates obtained online. In the closely related Chap. 7, a similar technique is used to solve the motion planning problem and design an adaptive servo-controller that is able to track time-varying references.

Alternatively, in Chap. 6, feedback input–output linearization is used together with a lumped model approximation of the DCSF infinite dimensional model, and adaptation is embedded using a control Lyapunov function technique.

Figure 8.1 shows the "location" of the different types of adaptive controllers considered in a "space" defined by performance and the degree of plant structure embedded in the plant model. As expected, experience shows that, by increasing the degree of physical structure embedded in the models that serve as a basis for controller design, performance increases, but the algorithm becomes more plant specific and may not be used, without significant changes, on plants other than the DCSF for which it is designed. For example, adaptive MPC algorithms like GPC or MUSMAR rely on data-driven linear models and may be used almost unchanged (requiring only a proper selection of the assumed model order and cost function weights) in plants other than DCSFs like a steam super-heater (Sect. 3.7.2) or arc welding (Sect. 3.7.3).

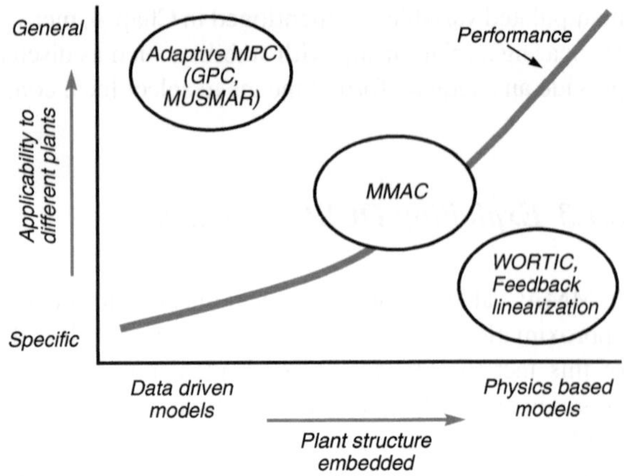

Fig. 8.1 Dependence on the degree of plant structure embedded in the model of performance and of the applicability of the adaptive control algorithms considered to different plants

At the other extreme, the WARTIC algorithms in Chap. 5 provide a better performance (they yield a step response with much less overshoot for the same raise time) than the adaptive MPC algorithms of Chap. 3, but may not be used directly on other plants (although the key idea of making a change of the time scale depending on the flow can be used with advantage).

Although not explicitly shown in Fig. 8.1, other algorithms that are important for applications are discussed in detail in the book, in relation to DCSFs. One of these topics is the use of flatness-based methods for planning variations of the outlet fluid temperature, addressed in Chap. 7. These type of methods are usually associated to robotic problems, but they provide valuable tools for solving servo problems for thermodynamic processes (i.e., tracking problems in which the reference is time varying) as well. For distributed parameter plants, their application involves a gauge type change of the time variable, which makes these methods intimately related to the WARTIC algorithms in Chap. 5.

8.1.4 Tackling the Adaptation Transient

Another example of a topic of central interest for applications is tackling the adaptation transient. Every adaptation algorithm entails an initial phase in which the estimates of the model parameters upon which the control decision relies are much deviated from the actual parameter values. This estimation error entails a transient in the plant that is beneficial to improve the estimates (the varying transient data is more informative), but that causes a loss in performance, (because the output temperature deviates significantly from the reference to track).

One possibility to prevent strong initial adaptation start-ups is to address the so-called dual control problem. The dual control problem, discussed in Sect. 3.9, consists of how to find an optimal balance between the dual actions of probing (system excitation in order to improve model estimation) and caution (reduction of excitation in order to improve control performance).

The exact solution of the dual control problem requires a computational load that renders it feasible only in very simple cases. As a consequence, even when controlling a DCSF with a sampling interval of more than 10 s, one has to resort to approximation methods in order to get a sub-optimal solution. As discussed in Sect. 3.9, a possibility to avoid undesirably strong start-up adaptation transients is to initialize the estimates of the parameters with values obtained from previous experience or, if not available, from simulations. It is remarked that the estimation error covariance matrix should also be initialized in a value that reflects the prior knowledge about the true value of the parameters. Another possibility is to approximate the exact solution of the dual control problem by solving a bi-criteria optimization problem (Sect. 3.9.1). The dual modification of MUSMAR, described in Sect. 3.9.2, leads to a smooth star-up in DCSFs with a small increase in the computational load, as shown by the results in Sect. 3.9.3.

8.1.5 Constraints

Finally, one should mention the way in which constraints are tackled. Given an estimate of the plant model parameters, the control algorithms considered in this book that optimize a cost compute the manipulated variable using an explicit formula. This is no longer possible if an hard constraint, that is to say a constraint on the exact value of the manipulated variable is imposed. Nevertheless, as discussed in Sect. 3.6.3, it is possible for some algorithms to impose a probabilistic constraint on the manipulated variable by adjusting the weight on the term of the cost associated to the manipulated variable. Since, close to a local minimum of the underlying cost, MUSMAR adjusts the controller gains along a Newton minimization direction, it is possible to prove that, if MUSMAR modified with a gain adjustment based on gradient rule converges, the only possible convergence points will be constrained minima. Other, more pedestrian, alternatives consist of adjusting the cost weights by trial and error or using a command governor that manipulates the reference such as to avoid fast changes.

Although no applications are reported in the literature for DCSS control, there are available a number of works that address the problem of incorporating constraints in adaptive MPC using different methods. See Sect. 3.5.2 for references.

8.2 The Context of Renewable Energy Production

The importance of control of DCSFs, and of adaptive control in particular, may perhaps be only fully understood by considering the wider context of renewable energy production. Of course, control is required to maintain energy production plants in a state that is adequate for their objectives, while ensuring equipment integrity

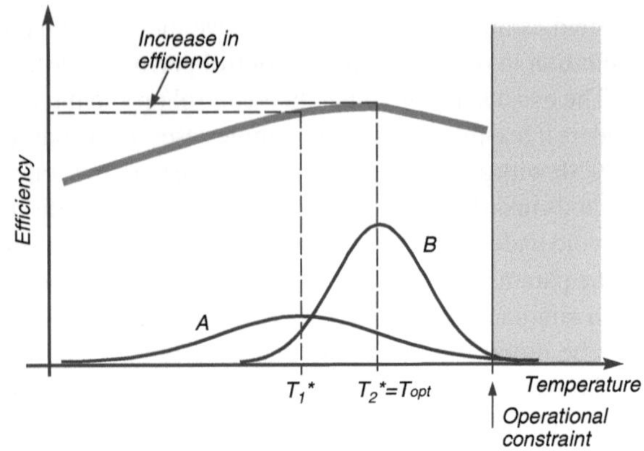

Fig. 8.2 Managing the outlet fluid temperature of a DCSF to maximize efficiency

and an extended useful life, i.e., enlarging the period in which the plant is able to operate in adequate conditions. However, the trend in energy markets raise new challenges that increasingly motivate the use of advanced control and optimization algorithms (Giannakis et al. 2013; Chakrabortty and Ilic 2012), and in particular of the ones that are able to cope with high levels of uncertainty, disturbances modeled as stochastic processes and time-varying references that, in the case of DCSFs cause in turn changes of linearized plant dynamics.

8.2.1 Efficiency and Operational Constraints

When considered in isolation, the objective of a DCSF is to convert as much solar power as possible to thermal energy that can be used, for instance, for energy production. As shown in Fig. 8.2, the fluid outlet temperature should be closed to the value T_{opt} that yields the maximum efficiency. However, due to the stochastic disturbances discussed in Sect. 1.3, the fluid temperature is not kept at an exact value, but instead randomly varies around its mean value with some probability density function (pdf). Let us call A to this pdf and T_1^* to the corresponding mean value (see Fig. 8.2).

For safety reasons, the percentage of samples of the temperature value that are allowed to be above a certain threshold is limited. This threshold corresponds to an operational constraint (shown in Fig. 8.2) and its value is selected depending on aspects such as the maximum temperature that the fluid to be heated tolerates or the possibility that the pipes and the other fluid containment equipments develop microcracks due to thermal stress. To comply with this constraint, the temperature set-point T_1^* is selected at a value that is lower than the optimum. This choice ensures that the probability of exceeding the operational constraint (given by the area under the pdf A and to the right of the constraint) is according to the specifications.

If a better control algorithm is used, for instance an adaptive controller that optimizes the performance in an asymptotic way, the fluctuations around the mean will be smaller. Therefore, as illustrated in Fig. 8.2, where B denotes the pdf in this new situation, the mean temperature T_2^* can be increased to become close to, or equal to, the optimum value, while maintaining the probability of exceeding the operational constraint. This aspect provides a direct link between the controller performance (measured by the ability of the controller to reject disturbances and compensate the DCSF dynamics so as to keep the outlet fluid temperature constant) and economic performance (measured by the power drawn from the plant).

In addition to efficiency, a better temperature regulation achievable by a controller with a higher performance has a positive impact on the quality of the energy being produced, in the sense that its physical qualities are closer to the desired specified values.

The operational constraint limit in Fig. 8.2 depends on the useful life considered in the plant design. If this upper bound is often exceeded by the temperature, micro-cracks start to appear in the pipes and in the other plant components that are subject to thermal stress. Ultimately, continuous grow of these micro-cracks will render the equipment unoperational. In life extending control (Kallappa 1997) the controller and the reference are tuned in such a way thjt the temperature fluctuates such as to maximize the plant useful life. The controllers designed under this objective may take advantage of material models (Ray and Tangirala 1996).

Although all the control strategies presented in this book have generally a beneficial effect in what concerns all the issues above, the methods of Chap. 3 have a tighter connection to their solution, due to the probabilistic formulation upon which they rely.

8.2.2 Market with Changing Prices

While classical energy markets are characterized by a strong regulation, with centralized power control being adequate, recent trends point in directions with increased degrees of freedom. An example is provided by distributed power grids with some degree of authority among local producers and energy markets with varying prices. Figure 8.3 illustrates a typical problem posed by new coming scenarios in relation to renewable energy production plants, including DCSFs.

As depicted in Fig. 8.3, at the present time t_p the plant dispatch manager accepts a contract that commits the plant to produce, between future times t_s and t_f a certain target value of power that should be inside a tolerance band. If the power produced at some instant falls outside the contracted tolerance band, a penalty must be paid. Fulfilling this contract implies the capacity to operate the DCSF using a controller that is able to keep the power produced close to the target value despite stochastic disturbances associated to incoming solar radiation and to incorporate predictive capacity in the control action, while taking into account the DCSF dynamics. This objective provides a motivation for using adaptive MPC controllers that are able to optimize a stochastic cost, such as the ones described in Chap. 3 or obtained by combining the algorithms in Chap. 3 with one of the exact linearization change of variables in either Chap. 5 or 6.

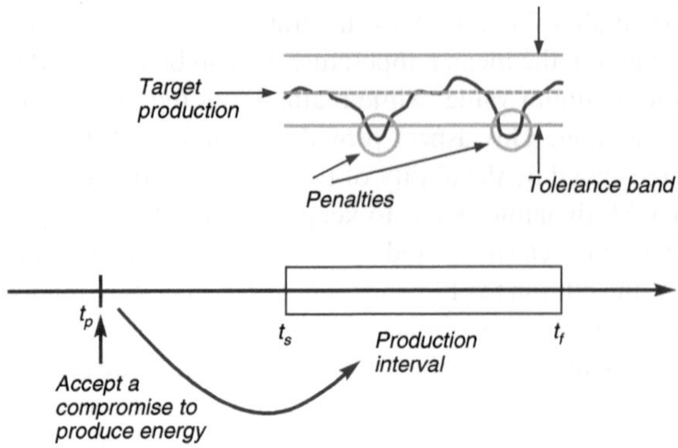

Fig. 8.3 Fulfilling a production contract in an uncertain framework

8.2.3 Cyber-Physical Systems

More involved problems reinforce the view that adaptive controllers such as the ones addressed in this book are basic "building blocks" of more complicated systems, such as networked systems connecting renewable energy sources that face uncertainty. The "green island" paradigm provides an example of such type of problems: Consider an island for which a number of power production and storage utilities are available. These utilities include DCSFs but also wind turbine fields and fossil fuel power generator units. The problem consists of taking advantage of these units such as to produce the electric power consumed by the island. DCSFs' production are subject to random variations due to incoming solar radiation. Wind turbines are also subject to random disturbances associated to wind speed. The problem is to compensate these fluctuations with storage and the fossil fuel unit production with the maximum efficiency. The total amount of power produced should match the island consumption that is also modeled by a stochastic process.

If the dispatch (i. e., the decision regarding which unit produces what amount of power and during which periods) is done in a centralized way, it can be structured as a hierarchical control problem in which a higher level block decides about the production of each unit and the controllers of the production units try to follow this reference. In both layers adaptive control can be used.

Another possibility is that there is no centralized dispatch and the units have to coordinate themselves to achieve the desired power production objective. This is the realm of distributed control that may also involve adaptive versions. Although other approaches are possible, the use of distributed MPC is imposed as a main streamline (Maestre and Negenborn 2014). An example of the use of the algorithms described in the present book in a distributed framework is the distributed version of SIORHC referred in Sect. 3.5.2 (Lemos and Igreja 2014).

Together with other modules, for instance for prediction and communication, the algorithms described in the present book may also be used as software bricks that can be interconnected in a structured way so as to design complex control systems, very much like using LEGO®[1] plastic blocks to build ingenious constructions. This use of the adaptive control algorithms described in this book is in the spirit of the emerging area of cyber-physical systems where computational and feedback control systems and communication networks are embedded with physical systems in a way that physical processes affect computations and vice versa (Lee and Seshia 2011; Kim and Kumar 2012). This framework provides a motivation to use and expand the methods addressed in this book in a wider context.

8.3 What Was Left Out?

A number of topics, although important for the matters treated above, have been left out. Some brief comments on them are therefore useful in order to establish what might be called the "boundary conditions" of this book.

The first topic is a more thorough treatment of control design by taking advantage of the fact that DCSF are bilinear systems, a class of nonlinear systems for which there is a rich plethora of results, for analysis as well as for controller design (Elliot 2009). Exploring possibilities offered by this family of methods, for which some work already exists for DCSFs (Carotenuto 1985, 1986), might be of interest to the more mathematically oriented reader.

Optimal Control (Speyer 2010; Geering 2007) provides also valuable tools that can be coupled with adaptation mechanisms. While nonlinear systems require the use of numerical methods (Betts 2010), early works address the control of DCSFs from an optimal point of view (Orbach 1981) and provide necessary optimality conditions for the manipulated variable. Since in both cases the control is computed by minimizing a cost, optimal control is intimately related with MPC. The approach taken in this book to solve the optimization problem associated to MPC consists of making a projection of the control function defined along the optimization horizon over a finite dimensional space, e.g., by making a piecewise constant approximation, and then solving the resulting finite dimensional optimization problem. An alternative would be to apply the necessary conditions of optimal control (Pontryagin's Maximum Principle in continuous time (Speyer 2010; Geering 2007) or its variant in discrete time (Goodwin et al. 2005)) and then, as in the previous case, proceed according to a receding horizon strategy. The approach based on optimal control has the advantage of imposing a structure on the control function that helps in the presence of local minima.

Stochastic optimization (Kall and Wallace 1994) is a tool used for planning on energy market problems (Garcia-Gonzalez 2008) but that can be used for real-time control as well, if applied according to a receding horizon strategy and in relation to the discussion in Sect. 8.2. Solving these problems involves the prediction of several types of variables (such as consumer's demand, solar radiation, wind speed) which is in itself an area of research.

---

[1] LEGO® is a registered trademark of the LEGO Group.

References

Anderson B, Brinsmead T, De Bruyne F, Hespanha J, Liberzon D, Morse AS (2000) Multiple model adaptive control. Part 1: Finite controller coverings. Int J Robust Nonlinear Control 10:909–929

Betts JT (2010) Practical methods for optimal control and estimation using nonlinear programming, 2nd edn. SIAM, Philadelphia

Carotenuto L, La Cava M, Raiconi G (1985) Regulator design for the bilinear distributed parameter of a solar power plant. Int J Sys Sci 16:885–900

Carotenuto L, La Cava M, Muraca P, Raiconi G (1986) Feedforward control for the distributed parameter model of a solar power plant. Large Scale Syst 11:233–241

Chakrabortty A, Ilic MD (eds) Control and optimization methods for electric smart grids. Springer, Berlin (2012)

Elliot DL (2009) Bilinear control systems. Springer, Berlin

Garcia-Gonzalez J, Muela R, Santosand LM, González AM (2008) Stochastic joint optimization of wind generation and pumped-storage units in an electricity market. IEEE Trans Power Syst 23(2):460–468

Geering HP (2007) Optimal control with engineering applications. Springer, Berlin

Giannakis GB, Kekatos V, Gatsis N, Kim S-J, Zhu H, Wollenberg BF (2013) Monitoring and optimization for power grids. IEEE Signal Proc Mag 30:108–128

Goodwin GC, Seron MM, De Dona JA (2005) Constrained control and estimation. Springer, Berlin

Kallappa P, Holmes M, Ray A (1997) Life extending control of fossil fuel power plants. Automatica 33(6):1101–1118

Kall P, Wallace SW (1994) Stochastic programming. Wiley, Chichester

Kim K-D, Kumar PR (2012) Cyber-physical systems: a perspective at the centennial. Proc IEEE 100:1287–1308

Lee EA, Seshia SA (2011) Introduction to embedded systems - a cyber-physical systems approach. http://LeeSeshia.org

Lemos JM, Igreja JM (2014) D-SIORHC, distributed MPC with stability constraints based on a game approach, Chapter 8. In: Maestre JM, Negenborn RR (eds) Distributed model predictive control made easy, Springer, Berlin, pp 133–146

Maestre JM, Negenborn RR (eds) (2014) Distributed model predictive control made easy. Springer, Berlin

Orbach A, Rorres C, Fischl R (1981) Optimal control of a solar collector loop using a distributed-lumped model. Automatica 27(3):535–539

Paoletti S, Juloski A Lj, Ferrari-Trecate G, Vidal R (2007) Identification of hybrid systems. Eur J Control 13(2–3):242–260

Ray A, Tangirala S (1996) Stochastic modelling of fatigue crack dynamics for on-line failure prognostics. IEEE Trans Autom Control 4(4):443–452

Speyer JL, Jacobson DH (2010) Primer on optimal control theory. SIAM, Philadelphia

Appendix A
Solution of the DCSF PDE

This appendix addresses the issue of solving the PDE (2.5).

A.1 Homogeneous Equation

Start by considering the homogeneous equation

$$\frac{\partial}{\partial t}T(z,\, t) + u(t)\frac{\partial}{\partial z}T(z,\, t) = 0, \tag{A.1}$$

to which the following ODE relating the independent variables t and x is associated

$$\frac{\mathrm{d}x}{\mathrm{d}t} - u(t) = 0. \tag{A.2}$$

The first integral of (A.2) is a relation of the form

$$\psi(x,\, t) = C$$

for C an arbitrary constant, satisfied by any solution $x = x(t)$ of (A.2), where the function ψ is not identically constant for all the values of x and t. In other words, the function ψ is constant along each solution of (A.2), with the constant C depending on the solution. Since in the case of equation (a1e1) there are 2 independent variables, there is only one functionally independent integrals Ibragimov (1999). By integrating (A.2) with respect to time, its first integral is found to be

$$\psi(x,\, t) = x - \int_{t_0}^{t} u(\sigma)\mathrm{d}\sigma. \tag{A.3}$$

J. M. Lemos et al., *Adaptive Control of Solar Energy Collector Systems*,
Advances in Industrial Control, DOI: 10.1007/978-3-319-06853-4,

A function $\psi(x, t)$ is a first integral of (A.2) if and only if it is a solution of the homogeneous PDE (A.1). Furthermore, or $t \geq t_0$, the general solution of (A.1) is given by

$$T(x,\ t) = F(\psi(x, t)), \tag{A.4}$$

where F is an arbitrary function Ibragimov (1999). The general solution of (A.1) is thus given by

$$T(x,\ t) = F(x - \int_{t_0}^{t} u(\sigma)d\sigma), \tag{A.5}$$

where $F(x)$ for $x \in [0,\ L]$ is the temperature distribution along the pipe at $t = t_0$.

□

A.2 Non-homogeneous Equation

Consider now the non-homogeneous Eq. (2.5) that, for the sake of simplifying the notation, is written as

$$\frac{\partial}{\partial t} T(z,\ t) = -u \frac{\partial}{\partial z} T(z,\ t) + g(x,\ t,\ T), \tag{A.6}$$

where

$$g(x,\ t,\ T) := \alpha R(t) - \gamma T(z,\ t). \tag{A.7}$$

To solve (A.6), and according to a procedure known as Laplace's method Ibragimov (1999), introduce the new set of coordinates $(\xi,\ \tau)$ given by

$$\xi = \varphi_1(x,\ t) := \psi(x, t) \tag{A.8}$$

and

$$\tau = \varphi_2(x,\ t) := t \tag{A.9}$$

where ψ is the first integral (A.3). The chain rule for derivatives yields

$$\frac{\partial T}{\partial x} = \frac{\partial \varphi_1}{\partial x} \cdot \frac{\partial T}{\partial \xi} + \frac{\partial \varphi_2}{\partial x} \cdot \frac{\partial T}{\partial \tau}, \tag{A.10}$$

$$\frac{\partial T}{\partial t} = \frac{\partial \varphi_1}{\partial t} \cdot \frac{\partial T}{\partial \xi} + \frac{\partial \varphi_2}{\partial t} \cdot \frac{\partial T}{\partial \tau}. \tag{A.11}$$

Since the transformation of variables is defined by (A.8, A.9), these equations reduce to

$$\frac{\partial T}{\partial x} = \frac{\partial \psi}{\partial x} \cdot \frac{\partial T}{\partial \xi}, \tag{A.12}$$

$$\frac{\partial T}{\partial t} = \frac{\partial \psi}{\partial t}\frac{\partial T}{\partial \xi} + \frac{\partial T}{\partial \tau}. \tag{A.13}$$

Therefore, in the new coordinates, and regrouping terms, the non-homogeneous equation (a1e7) is written

$$\frac{\partial T}{\partial \tau} + \left[\frac{\partial \psi}{\partial t} + u\frac{\partial \psi}{\partial x}\right]\frac{\partial \psi}{\partial \xi} = g. \tag{A.14}$$

Since ψ satisfies the homogeneous equation (A.1) and $\tau = t$, this equation reduces to the ODE

$$\frac{\partial T}{\partial t} = g \tag{A.15}$$

where x is to be expressed in terms of t and ξ by solving (A.8) with respect to x. Since g is given by (A.7), the above ODE reads

$$\frac{\partial}{\partial t}T(\xi, t) = -\gamma T(\xi, t) + \alpha R(t). \tag{A.16}$$

Equation (A.16) is a linear, scalar, first order ODE whose solution is

$$T(\xi, t) = T(\xi, t_0)e^{-\gamma(t-t_0)} + \alpha \int_{t_0}^{t} R(\sigma)e^{\gamma(\sigma-t)}d\sigma. \tag{A.17}$$

Inverting the change of variable (A.8) and using (A.3) yields (2.6). □

Since the transformation of variables is defined by (A.8, A.9), these equations reduce to

[illegible]

[illegible]

Appendix B
Recursive Least Squares Deduction

In order to deduce Eqs. (3.73)–(3.75) start by considering the following lemma:

Matrix inversion lemma

For A, B, C and D matrices of convenient dimensions such that the indicated inversions exist, it holds that

$$[A + BCD]^{-1} = A^{-1} - A^{-1}\left[DA^{-1}B + C^{-1}\right]^{-1} DA^{-1}. \qquad \text{(B.1)}$$

Proof Right multiply the right hand side of (B.1) by $A + BCD$ to get

$$\begin{aligned}
&\left(A^{-1} - A^{-1}\left[DA^{-1}B + C^{-1}\right]^{-1} DA^{-1}\right)(A + BCD) \\
&\quad = I - A^{-1}B\left[DA^{-1}B + C^{-1}\right]^{-1} D \\
&\qquad + A^{-1}BCD - A^{-1}B\left[DA^{-1}B + C^{-1}\right]^{-1} DA^{-1}BCD \\
&\quad = I + A^{-1}B\left[DA^{-1}B + C^{-1}\right]^{-1}\left\{\left[DA^{-1}B + C^{-1}\right]CD - D - DA^{-1}BCD\right\} \\
&\quad = I.
\end{aligned}$$

Now, left multiply the right hand side of (B.1) by $A + BCD$ to get

$$\begin{aligned}
(A + BCD)&\left(A^{-1} - A^{-1}\left[DA^{-1}B + C^{-1}\right]^{-1} DA^{-1}\right) \\
&= I - B\left[DA^{-1}B + C^{-1}\right]^{-1} DA + BCDA^{-1} \\
&\quad - BCDA^{-1}B\left[DA^{-1}B + C^{-1}\right]^{-1} DA \\
&= I + \left\{-B + BC\left[DA^{-1}B + C^{-1}\right] - BCDA^{-1}B\right\}
\end{aligned}$$

J. M. Lemos et al., *Adaptive Control of Solar Energy Collector Systems*,
Advances in Industrial Control, DOI: 10.1007/978-3-319-06853-4,

$$\cdot \left[DA^{-1}B + C^{-1} \right] DA = I$$

□

Let now

$$\bar{z} \triangleq \begin{bmatrix} z(1) \\ \vdots \\ z(k) \end{bmatrix}, \tag{B.2}$$

$$\Phi_{LS} \triangleq \begin{bmatrix} \varphi^T(1) \\ \vdots \\ \varphi^T(k) \end{bmatrix} \tag{B.3}$$

and

$$\bar{v} \triangleq \begin{bmatrix} v(1) \\ \vdots \\ v(k) \end{bmatrix}. \tag{B.4}$$

These quantities are related by the matrix regression model written for all the available data

$$\bar{z} = \Phi_{LS}\vartheta + \bar{v}. \tag{B.5}$$

With this notation the least squares functional (3.72) can be written

$$J_{LS}(\vartheta) = \frac{1}{2} \left(\bar{z} - \Phi_{LS}\vartheta\right)^T M_{LS} \left(\bar{z} - \Phi_{LS}\vartheta\right), \tag{B.6}$$

where $M_{LS} \in \mathbb{R}^{k \times k}$ is the diagonal matrix of weights

$$M_{LS} \triangleq \begin{bmatrix} \lambda^k & 0 & \dots & 0 \\ 0 & \lambda^{k-1} & \ddots & \vdots \\ \vdots & \ddots & \ddots & 0 \\ 0 & \dots & 0 & \lambda \end{bmatrix}. \tag{B.7}$$

Equating the gradient of $J_{LS}(\vartheta)$ with respect to ϑ to zero yields the normal equation satisfied by the least squares estimates $\hat{\vartheta}$ of ϑ

$$\Lambda(k)\hat{\vartheta}(k) = \Phi_{LS}^T M_{LS}\bar{z}, \tag{B.8}$$

where $\Lambda(k)$ is the information matrix given k observations, given by

$$\Lambda(k) = \Phi_{LS}^T M_{LS}\Phi_{LS}. \tag{B.9}$$

If the experimental conditions are such that the data verify a persistency of excitation condition, then the inverse of $\Lambda(k)$ exists, and the least squares estimates are given by

$$\hat{\vartheta}(k) = \Lambda^{-1}(k)\Phi_{LS}^T M_{LS}\bar{z}. \tag{B.10}$$

This expression can be given the form

$$\hat{\vartheta}(k) = \Lambda^{-1}(k)\sum_{i=1}^{k}\lambda^{k-i}\varphi(i)z(i). \tag{B.11}$$

Furthermore, the information matrix verifies the recursive equation

$$\Lambda(k) = \lambda\Lambda(k-1) + \varphi(k)\varphi^T(k). \tag{B.12}$$

Isolate the last term of the summation in (B.11) and use (B.11) with k replaced by $k-1$ to get

$$\hat{\vartheta}(k) = \Lambda^{-1}(k)\left\{\varphi(k)y(k) + \lambda\Lambda(k-1)\hat{\vartheta}(k-1)\right\} \tag{B.13}$$

and then express $\Lambda(k-1)$ in terms of $\Lambda(k)$ using (B.12), yielding

$$\hat{\vartheta}(k) = \hat{\vartheta}(k-1) + \Lambda^{-1}(k)\varphi(k)\left[y(k) - \varphi^T(k)\hat{\vartheta}(k-1)\right]. \tag{B.14}$$

Together with (B.12), Eq. (B.14) provides a way of updating recursively the estimate of ϑ. The estimate given the data up to time k is obtained by correcting the estimate given the observations up to time $k-1$ by a term given by the product of a gain (the Kalman gain, $\Lambda^{-1}(k)\varphi(k)$) by the a priori prediction error $y(k) - \varphi^T(k)\hat{\vartheta}(k-1)$. These expressions have the drawback of requiring the computation of the inverse of the information matrix Λ at each iteration. This can be avoided by propagating in time directly the inverse of Λ. Let

$$P(k) = \Lambda^{-1}(k). \tag{B.15}$$

Matrix P is proportional to the parameter error covariance error and, for simplicity, is referred simply as the "covariance matrix". From (B.12) it follows that

$$P(k) = \lambda^{-1}\left[\Lambda(k-1) + \varphi(k)\lambda^{-1}\varphi^T(k)\right]^{-1}. \tag{B.16}$$

Using the matrix inversion lemma (B.1) with $A = \Lambda(k-1)$, $B = \varphi(k)$, $C = \lambda^{-1}$ and $D = \varphi^T(k)$ yields (3.75). Equation (3.73) follows from (B.14) and the definition of P. □

Appendix C
MUSMAR Models

In this appendix we explain how the predictive models used by MUSMAR, (3.62) are obtained from the ARX model (3.9). As explained in Sect. 3.2.4 the MUSMAR adaptive control algorithm restricts the future control samples (with respect to the present discrete time denoted k), from time $k+1$ up to $t+T-1$, to be given by a *constant feedback* of the pseudo-state, leaving $u(k)$ free. In order to see how the predictive model (3.47) is modified by this assumption, start by observing that the pseudostate $s(k)$ defined in (3.95) satisfies the dynamic state equation

$$s(k+1) = \Phi_s s(k) + \Gamma_s u(k) + \bar{e}_1^T e(k), \tag{C.1}$$

in which

$$\bar{e}_1^T \triangleq [1\,0\,\ldots\,0], \tag{C.2}$$

$$\Phi_s \triangleq \begin{bmatrix} \psi_1^T \\ I_{n-1}\;\underline{0}_{(n-1)\times n} \\ \underline{0}_{1\times(n+m)} \\ \underline{0}_{(m-1)\times n}\;I_{m-1}\;\underline{0}_{(m-1)\times 1} \end{bmatrix}, \tag{C.3}$$

$$\Gamma_s \triangleq [b_0\,0\,\ldots\,0\,1\,0\,\ldots\,0]^T, \tag{C.4}$$

and

$$\psi_1^T \triangleq [-a_1\;\ldots\;-a_n b_1\;\ldots\;b_m]. \tag{C.5}$$

The matrix entries a_i and b_j above are the coefficients of the ARX model (3.9) and e is the innovations sequence of the same model.

Assume that a constant feedback law written as

$$u(k+i) = F_0^T s(k+i) \tag{C.6}$$

J. M. Lemos et al., *Adaptive Control of Solar Energy Collector Systems*, Advances in Industrial Control, DOI: 10.1007/978-3-319-06853-4,

will be acting on the plant from time $k+1$ up to time $t+T-1$, where T is the optimization horizon (do not confuse the horizon T with similar symbol used to refer the transpose of a matrix; the context in which they are used should be sufficient to distinguish both). Under this assumption, from time $k+1$ up to time $k+T-1$, the pseudo-state verifies

$$s(k+i+1) = \Phi_F s(k+i) + \bar{e}_1 e(k+i), \tag{C.7}$$

with

$$\Phi_F \triangleq \Phi_s + \Gamma_s F_0^T. \tag{C.8}$$

Combining (C.1) and (C.7), it is concluded by making a comparison with Kalman predictors Goodwin and Sin (1984) that the optimal predictor of $s(k+i)$, given information up to time k is

$$\hat{s}(k+i|k) = \Phi_F^{i-1}\left[\Phi_s s(k) + \Gamma u(k)\right]. \tag{C.9}$$

The optimal predictor for $u(k+i)$ is

$$\hat{u}(k+i|k) = F_0^T \hat{s}(k+i|k), \tag{C.10}$$

which implies that

$$\hat{u}(k+i|k) = F_0^T \Phi_F^{i-1}\left[\Phi_s s(k) + \Gamma u(k)\right]. \tag{C.11}$$

Defining

$$\mu_i \triangleq F_0^T \Phi_F^{i-1}\Gamma_s \tag{C.12}$$

and

$$\phi_i^T \triangleq F_0^T \Phi_F^{i-1}\Phi_s, \tag{C.13}$$

the expression (C.11) reduces to the MUSMAR input predictor (3.63).

Furthermore, since by the construction of the pseudo-state the optimal predictor for the output $y(k+i)$, given observations up to time k, is

$$\hat{y}(k+i|k) = \bar{e}_1^T \hat{s}(k+i|k), \tag{C.14}$$

it follows that

$$\hat{y}(k+i|k) = \bar{e}_1^T \Phi_F^{i-1}\left[\Phi_s s(k) + \Gamma u(k)\right]. \tag{C.15}$$

Defining

$$\theta_i \triangleq \bar{e}_1^T \Phi_F^{i-1}\Gamma_s \tag{C.16}$$

and

$$\psi_i^T \triangleq \bar{e}_1^T \Phi_F^{i-1} \Phi_s, \tag{C.17}$$

it is also recognized that the expression (C.15) reduces to the MUSMAR output predictor (3.62). □

Appendix D
MUSMAR as a Newton Minimization Algorithm

In Mosca et al. (1989) Propositions 1 and 2 of Chap. 3 that characterize the possible convergence points of MUSMAR and the direction it yields for the progression of the controller gains are proved using the ODE method for studying the convergence of stochastic algorithms. In this appendix we use a different method that has the advantage of not relying on the ODE method, being thus more comprehensive.

D.1 Proof of Proposition 1

According to the control strategy used

$$\tilde{y}(t+j) \approx \theta_j(F_{k-1}) + H_j'(F_{k-1},\ F_k)s(t) \tag{D.1}$$

and

$$u(t+j-1) \approx \mu_{j-1}(F_{k-1})\eta(t) + G_{j-1}'(F_{k-1},\ F_k)s(t), \tag{D.2}$$

where

$$H_j(F_{k-1},\ F_k) = \psi_j(F_{k-1}) + \theta_j(F_{k-1}F_k) \tag{D.3}$$

and

$$G_{j-1}(F_{k-1},\ F_k) = \phi_{i-1}(F_{k-1})F_k \tag{D.4}$$

Let F_k be computed according to (3.104, 3.105). By adding and subtracting $\theta_j^2(F_{k-1}F_{k-1})$ and $\mu_i^2(F_{k-1})F_{k-1}\rho$, this becomes

J. M. Lemos et al., *Adaptive Control of Solar Energy Collector Systems*,
Advances in Industrial Control, DOI: 10.1007/978-3-319-06853-4,

$$F_k = -\frac{1}{\alpha(F_{k-1})}\left(\sum_{j=1}^{T}\theta_j(F_{k-1})H_j(F_{k-1}, F_{k-1}) + \rho\sum_{j=1}^{T-1}\mu_j(F_{k-1})G_j(F_{k-1}, F_{k-1})\right)$$
$$+\frac{F_{k-1}}{\alpha(F_{k-1})}\left(\sum_{j=1}^{T}\theta_j^2(F_{k-1}) + \rho\sum_{j=1}^{T-1}\mu_i^2(F_{k-1})\right). \tag{D.5}$$

If $\rho F_{k-1}/\alpha(F_{k-1})$ is added and subtracted, (D.5) becomes

$$F_k = -\frac{1}{\alpha(F_{k-1})}\left(\sum_{j=1}^{T}\theta_j(F_{k-1})H_j(F_{k-1}, F_{k-1}) + \rho\sum_{j=1}^{T-1}\mu_j(F_{k-1})G_j(F_{k-1}, F_{k-1}) + \rho F_{k-1}\right)$$
$$+\frac{F_{k-1}}{\alpha(F_{k-1})}\left(\sum_{j=1}^{T}\theta_j^2(F_{k-1}) + \rho\left(1+\sum_{j=1}^{T-1}\mu_i^2(F_{k-1})\right)\right). \tag{D.6}$$

The gradient $\nabla_T J(F)$ of the receding horizon cost $J_T(t)$ with respect to the controller gains is given by

$$\nabla_T J(F) = E\left[\sum_{j=1}^{T}\tilde{y}(t+j)\frac{\partial\tilde{y}(t+j)}{\partial F} + \rho u(t+j-1)\frac{\partial u(t+j-1)}{\partial F}\right]$$
$$= \sum_{j=1}^{T}(\theta_i(F_{k-1})E[y(t+j)s(t)] + \rho\mu_{j-1}(F_{k-1})E[u(t+j-1)s(t)]) \tag{D.7}$$

or, equivalently, by

$$\frac{1}{2}R_s^{-1}\nabla_T J(F) = \sum_{i=1}^{T}(\theta_i(F_{k-1})H_i(F_{k-1}, F_k)$$
$$+\rho\mu_{j-1}(F_{k-1})G_{j-1}(F_{k-1}, F_k)). \tag{D.8}$$

Here, (D.3, D.4) and (3.105) are used together with

$$E[y(t+j)s(t)] = R_s H_j(F_{k-1}, F_k) \tag{D.9}$$

and

$$E[u(t+j)s(t)] = R_s G_j(F_{k-1}, F_k). \tag{D.10}$$

Since $\mu_0 = 1$, $\phi_0 = 1$ and $G_0(F_{k-1}, F_{k-1})$, the conclusion follows from (D.8) and (D.6).

□

D.2 Proof of Proposition 2

Let iterations (3.110) be initialized in a neighborhood of a local minimum F^* of J_∞, small enough so that J_∞ has a single isolated minimum in it. In this region define the candidate Lyapunov function

$$V(F) = J_\infty(F) - J_\infty(F^*). \tag{D.11}$$

This function is continuous and differentiable with respect to F and vanishes for $F = F^*$, at which point it has a local minimum. Furthermore, it is strictly decreasing around the trajectories of (3.110). Indeed

$$\begin{aligned} V(F_k) - V(F_{k-1}) &= J_\infty(F_k) - J_\infty(F_{k-1}) \\ &= (\nabla J_\infty(F_{k-1}))' \cdot (F_k - F_{k-1}) + o(F_k - F^*) \\ &= -\frac{1}{\alpha(F_{k-1})} (\nabla J_\infty(F_{k-1}))' R_s^{-1}(F_{k-1}) \nabla J_\infty(F_{k-1}) \\ &\quad + o(F_k - F^*). \end{aligned} \tag{D.12}$$

Hence, and since it can be proved that $R_s > 0$ Mosca et al. (1989), in a sufficiently small neighborhood of F^*,

$$V(F-k) - V(F_{k-1}) < 0 \tag{D.13}$$

and by Lyapunov's Direct Method the sequence F_k converges to F^*.

In an analogous way, around a local maximum $F_{\max}$ one can define the function

$$W(F) = -J_\infty(F) + J_\infty(F_{\max}). \tag{D.14}$$

This function is continuous and differentiable and has a local minimum at $F_{\max}$ but now, in a neighborhood of the maximum, it may be proved using similar arguments as above that (3.110) yields

$$W(F_k) - W(F_{k-1}) > 0. \tag{D.15}$$

This inequality allows to conclude that $F_{\max}$ is unstable in the sense of Lyapunov. Hence, the recursion (3.110) does not yield a sequence of gains that converge to a maximum.

□

Appendix E
Derivation of the MUSMAR Dual Algorithm

For the sake of simplifying the notation, the mean conditioned on the observations $E\left\{\cdot|O^k\right\}$ is denoted by $E\left\{\cdot\right\}$.

E.1 Deduction of Equation (3.137)

Since,

$$\begin{aligned}
E\left\{y^2(k+i)\right\} &= E\left\{\left[\theta_i u(k) + \psi_i^T s(k)\right]^2\right\} \\
&= E\left\{\theta_i^2 u^2(k) + 2u(k)\theta_i \psi_i^T s(k) + s^T(k)\psi_i \psi_i^T s(k)\right\} \\
&= E\left\{\theta_i^2\right\} u^2(k) + 2u(k) E\left\{\theta_i \psi_i^T\right\} s(k) + s^T(k) E\left\{\psi_i \psi_i^T\right\} s(k) \\
&= \left[\hat{\theta}_i^2 + \sigma_{\theta i}^2\right] u^2(k) + 2u(k)\left[\hat{\theta}_i \hat{\psi}_i^T + \sigma_{\theta\psi i}^T\right] s(k) \\
&\quad + s^T(k) E\left\{\psi_i \psi_i^T\right\} s(k)
\end{aligned} \tag{E.1}$$

and

$$\begin{aligned}
E\left\{u^2(k+i-1)\right\} &= E\left\{\left[\mu_{i-1} u(k) + \phi_{i-1}^T s(k)\right]^2\right\} \\
&= E\left\{\mu_{i-1}^2 u^2(k) + 2u(k)\mu_{i-1}\phi_{i-1}^T s(k) + s^T(k)\phi_{i-1}\phi_{i-1}^T s(k)\right\} \\
&= E\left\{\mu_{i-1}^2\right\} u^2(k) + 2u(k) E\left\{\mu_{i-1}\phi_{i-1}^T\right\} s(k) + s^T(k) E\left\{\phi_{i-1}\phi_{i-1}^T\right\} s(k) \\
&= \left[\hat{\mu}_{i-1}^2 + \sigma_{\mu i}^2\right] u^2(k) + 2u(k)\left[\hat{\mu}_{i-1}\hat{\phi}_{i-1}^T + \sigma_{\mu\phi i}^T\right] s(k) \\
&\quad + s^T(k) E\left\{\phi_{i-1}\phi_{i-1}^T\right\} s(k)
\end{aligned} \tag{E.2}$$

J. M. Lemos et al., *Adaptive Control of Solar Energy Collector Systems*, Advances in Industrial Control, DOI: 10.1007/978-3-319-06853-4,

the minimization of the cost function according to Eq. (3.133) leads to

$$
\begin{aligned}
\frac{T}{2}\frac{\partial J_k^c}{\partial u(k)} &= \frac{1}{2}\frac{\partial}{\partial u(k)}\left(\sum_{i=1}^{T}\left[E\left\{y^2(k+i)\right\}+\rho E\left\{u^2(k+i-1)\right\}\right]\right) \\
&= \left[\sum_{i=1}^{T}\hat{\theta}_i^2+\sigma_{\theta i}^2+\rho\hat{\mu}_{i-1}^2+\rho\sigma_{\mu i}^2\right]u(k) \\
&\quad+\left[\sum_{i=1}^{T}\hat{\theta}_i\hat{\psi}_i^T+\sigma_{\theta\psi i}^T+\rho\hat{\mu}_{i-1}\hat{\phi}_{i-1}^T+\rho\sigma_{\mu\phi i}^T\right]s(k)=0
\end{aligned}
\tag{E.3}
$$

or

$$
u_c(k)=-\frac{\sum_{i=1}^{T}\hat{\theta}_i\hat{\psi}_i^T+\rho\sum_{i=1}^{T}\hat{\mu}_{i-1}\hat{\phi}_{i-1}^T+\sum_{i=1}^{T}(\sigma_{\theta\psi i}^T+\rho\sigma_{\mu\phi i}^T)}{\sum_{i=1}^{T}\hat{\theta}_i^2+\rho\sum_{i=1}^{T}\hat{\mu}_{i-1}^2+\sum_{i=1}^{T}(\sigma_{\theta i}^2+\rho\sigma_{\mu i}^2)}s(k). \tag{E.4}
$$

Since the data vector, $z(k)=\left[u(k)\, s^T(k)\right]^T$, used to estimate the predictive models parameters is common to all models (actually, the vector used is $z(k-T)$ so the T-steps ahead predictor can use the last output available to perform the estimation and that is $y(k)$), there is only one RLS covariance matrix, $P(k)$ common to all models. To obtain the parameters of the true covariance matrix, matrix $P(k)$ should be multiplied by the variance of the prediction error of each model. Therefore,

$$
\begin{array}{l}
\sigma_{\theta i}^2=p_{uu}\sigma_{yi}^2;\ \sigma_{\theta\psi i}=p_{us}\sigma_{yi}^2; \\
\sigma_{\mu i}^2=p_{uu}\sigma_{ui}^2;\ \sigma_{\mu\phi i}=p_{us}\sigma_{ui}^2;
\end{array},
\tag{E.5}
$$

with

$$
P=\begin{bmatrix} p_{uu} & p_{us}^T \\ p_{us} & p_{ss}\end{bmatrix}, \tag{E.6}
$$

where the argument k is omitted for notation simplicity, which yields the control law

$$
u_c(k)=-\frac{\sum_{i=1}^{T}\hat{\theta}_i\hat{\psi}_i^T+\rho\sum_{i=1}^{T}\hat{\mu}_{i-1}\hat{\phi}_{i-1}^T+\beta p_{us}^T}{\sum_{i=1}^{T}\hat{\theta}_i^2+\rho\sum_{i=1}^{T}\hat{\mu}_{i-1}^2+\beta p_{uu}}s(k). \tag{E.7}
$$

□

E.2 Deduction of Equation (3.140)

Since J_k^a is quadratic with negative second derivative, the minimum of this criteria coincides with one of the borders of the interval, and is given by

$$u(k) = u_c(k) + \vartheta(k)\text{sgn}\left\{J_k^a(u_c(k) - \vartheta(k)) - J_k^a(u_c(k) + \vartheta(k))\right\} \tag{E.8}$$

The expected prediction error for the output and input can be computed as

$$\begin{aligned}
E\left\{\left[y(k+i) - \hat{y}(k+i)\right]^2\right\} &= E\left\{\left[(\theta_i - \hat{\theta}_i)u(k) + (\psi_i^T - \hat{\psi}_i^T)s(k)\right]^2\right\} \\
&= E\left\{(\theta_i - \hat{\theta}_i)^2\right\}u^2(k) + 2u(k)E\left\{(\theta_i - \hat{\theta}_i)(\psi_i^T - \hat{\psi}_i^T)\right\}s(k) \\
&\quad + s^T(k)E\left\{(\psi_i - \hat{\psi}_i)(\psi_i^T - \hat{\psi}_i^T)\right\}s(k) \\
&= \sigma_{\theta i}^2 u^2(k) + 2u(k)\sigma_{\theta\psi i}^T s(k) \\
&\quad + s^T(k)E\left\{(\psi_i - \hat{\psi}_i)(\psi_i^T - \hat{\psi}_i^T)\right\}s(k),
\end{aligned} \tag{E.9}$$

$$\begin{aligned}
E\left\{\left[u(k+i-1) - \hat{u}(k+i-1)\right]^2\right\} &= E\left\{\left[(\mu_{i-1} - \hat{\mu}_{i-1})u(k) + (\phi_{i-1}^T - \hat{\phi}_{i-1}^T)s(k)\right]^2\right\} \\
= E\left\{(\mu_{i-1} - \hat{\mu}_{i-1})^2\right\}u^2(k) &+ 2u(k)E\left\{(\mu_{i-1} - \hat{\mu}_{i-1})(\phi_{i-1}^T - \hat{\phi}_{i-1}^T)\right\}s(k) \\
+ s^T(k)E\left\{(\phi_{i-1} - \hat{\phi}_{i-1})(\phi_{i-1}^T - \hat{\phi}_{i-1}^T)\right\}s(k) \\
= \sigma_{\mu i}^2 u^2(k) + 2u(k)\sigma_{\mu\phi i}^T s(k) &+ s^T(k)E\left\{(\phi_{i-1} - \hat{\phi}_{i-1})(\phi_{i-1}^T - \hat{\phi}_{i-1}^T)\right\}s(k).
\end{aligned} \tag{E.10}$$

Replacing (E.9) and (E.10) in

$$\begin{aligned}
J_k^a &= -E\left\{\frac{1}{T}\sum_{i=1}^{T}\left[y(k+i) - \hat{y}(k+i)\right]^2 + \alpha\left[u(k+i-1) - \hat{u}(k+i-1)\right]^2\right\} \\
&= -\frac{1}{T}\sum_{i=1}^{T}E\left\{\left[y(k+i) - \hat{y}(k+i)\right]^2\right\} \\
&\quad - \frac{\alpha}{T}\sum_{i=1}^{T}E\left\{\left[u(k+i-1) - \hat{u}(k+i-1)\right]^2\right\}
\end{aligned} \tag{E.11}$$

yields

$$J_k^a(u) = -\frac{1}{T}\sum_{i=1}^{T}\left[\sigma_{\theta i}^2 + \alpha\sigma_{\mu i}^2\right]u^2(k) + 2u(k)\left[\sigma_{\theta\psi i}^T + \alpha\sigma_{\mu\phi i}^T\right]s(k)$$

$$\underbrace{-\frac{1}{T}\sum_{i=1}^{T}s^T(k)\left[E\left\{(\psi_i - \hat{\psi}_i)(\psi_i^T - \hat{\psi}_i^T)\right\} + \alpha s^T(k)E\left\{(\phi_{i-1} - \hat{\phi}_{i-1})(\phi_i^T - \hat{\phi}_{i-1}^T)\right\}\right]s(k)}_{\bar{c}}$$

$$= - \underbrace{\left[\frac{1}{T} \sum_{i=1}^{T} (\sigma_{yi}^2 + \alpha \sigma_{ui}^2) \right]}_{\gamma \geq 0} \left[p_{uu} u^2(k) + 2u(k) p_{us}^T s(k) \right] + \bar{c} \tag{E.12}$$

and

$$\begin{aligned} J_k^a(u_c(k) - \vartheta(k)) - J_k^a(u_c(k) + \vartheta(k)) & \\ = - \gamma \left[p_{uu} (u_c^2(k) - 2u_c(k)\vartheta(k) + \vartheta(k)^2) + 2(u_c(k) - \vartheta(k)) p_{us}^T s(k) \right] & \\ + \gamma \left[p_{uu} (u_c^2(k) + 2u_c(k)\vartheta(k) + \vartheta(k)^2) + 2(u_c(k) + \vartheta(k)) p_{us}^T s(k) \right] & \\ = 4\gamma \vartheta(k) \left[p_{uu} u_c(k) + p_{us}^T s(k) \right]. & \end{aligned} \tag{E.13}$$

Equation (E.8) reduces then to Eq. (3.140). □

E.3 Local Models and Local Controllers for the Air Heating Fan Example

This appendix contains the numeric data for SMMAC control of the air heating fan laboratory plant, as described in Sect. 4.3. The local models are described in Table E.1 and the local controllers, that match each of them, are described in Table E.2.

Table E.1 Estimated local models of the air heating fan plant

	Low temperature	Medium temperature
Low flow	$A^*(q^{-1}) = 1 - 1.84q^{-1} + 0.8442q^{-2}$	$A^*(q^{-1}) = 1 - 1.862q^{-1} + 0.8649q^{-2}$
	$B^*(q^{-1}) = 0.001866q^{-3} - 0.0003835q^{-4}$	$B^*(q^{-1}) = 0.003823q^{-2}$
	$C^*(q^{-1}) = 1 - 1.389q^{-1} + 0.5118q^{-2}$	$C^*(q^{-1}) = 1 - 1.15q^{-1} + 0.3074q^{-2}$
Medium flow	$A^*(q^{-1}) = 1 - 1.82q^{-1} + 0.8257q^{-2}$	$A^*(q^{-1}) = 1 - 1.83q^{-1} + 0.8348q^{-2}$
	$B^*(q^{-1}) = 0.00143q^{-2}$	$B^*(q^{-1}) = 0.003763q^{-2}$
	$C^*(q^{-1}) = 1 - 1.349q^{-1} + 0.4493q^{-2}$	$C^*(q^{-1}) = 1 - 1.118q^{-1} + 0.3484q^{-2}$
High flow	$A^*(q^{-1}) = 1 - 1.783q^{-1} + 0.7902q^{-2}$	$A^*(q^{-1}) = 1 - 1.8q^{-1} + 0.806q^{-2}$
	$B^*(q^{-1}) = 0.001222q^{-2} + 0.0002227q^{-3}$	$B^*(q^{-1}) = 0.00399q^{-2}$
	$C^*(q^{-1}) = 1 - 1.334q^{-1} + 0.4904q^{-2}$	$C^*(q^{-1}) = 1 - 1.193q^{-1} + 0.4593q^{-2}$

Table E.2 Local controllers for the different operating regions of the air heating fan plant

	Low temperature	Medium temperature
Low flow	$R^*(q^{-1}) = 1 - 1.696q^{-1} + 0.8719q^{-2}$ $-0.1910q^{-3} + 0.01537q^{-4}$	$R^*(q^{-1}) = 1 - 1.156q^{-1} + 0.1556q^{-2}$
	$S^*(q^{-1}) = 37.02 - 70.75q^{-1} + 33.84q^{-2}$	$S^*(q^{-1}) = 41.41 - 76.24q^{-1} + 35.18q^{-2}$
	$T^*(q^{-1}) = 0.8677q^{-1} - 1.205q^{-2}$ $+0.4440q^{-3}$	$T^*(q^{-1}) = 2.176 - 2.503q^{-1} + 0.6690q^{-2}$
Medium flow	$R^*(q^{-1}) = 1 - 1.471q^{-1} + 0.4707q^{-2}$	$R^*(q^{-1}) = 1 - 1.170q^{-1} + 0.1698q^{-2}$
	$S^*(q^{-1}) = 27.47 - 52.17q^{-1} + 24.81q^{-2}$	$S^*(q^{-1}) = 51.96 - 96.19q^{-1} + 44.63q^{-2}$
	$T^*(q^{-1}) = 1.132 - 1.526q^{-1} + 0.5086q^{-2}$	$T^*(q^{-1}) = 1.742 - 1.947q^{-1} + 0.6069q^{-2}$
High flow	$R^*(q^{-1}) = 1 - 1.479q^{-1} + 0.4935q^{-2}$ $-0.01485q^{-3}$	$R^*(q^{-1}) = 1 - 1.291q^{-1} + 0.2912q^{-2}$
	$S^*(q^{-1}) = 58.66 - 111.1q^{-1} + 52.68q^{-2}$	$S^*(q^{-1}) = 51.48 - 96.22q^{-1} + 45.07q^{-2}$
	$T^*(q^{-1}) = 1.756 - 2.342q^{-1} + 0.8611q^{-2}$	$T^*(q^{-1}) = 1.254 - 1.497q^{-1} + 0.5760q^{-2}$

Appendix F
Warped Time Optimization

In this appendix the underlying control laws (i.e., the control law assuming that the plant parameters are known) used in the WARTIC-i/o and WARTIC-state control algorithms, described in Chap. 5, are deduced.

F.1 Proof of the WARTIC-i/o Control Law

Hereafter we prove Eq. (5.16) that yields a closed-form expression for the WARTIC-i/o control law. For that sake, consider the predictive model (5.15) and assume that u is constant and equal to $u(k)$ over the prediction horizon, yielding

$$T_0(k+i) = \alpha u(k) \sum_{j=1}^{i} R(k-1+j) + \alpha \sum_{p=1}^{n-i} R(k-p)u(k-p) + \beta T_{in}(k+i-n). \tag{F.1}$$

Assume now that the future values of radiation at time $k+1$ up to time $k+T$ (that are unknown at time k) are equal to $R(k)$. Equation (F.1) becomes

$$T_0(k+i) = \alpha u(k)R(k)i + \alpha \sum_{p=1}^{n-i} R(k-p)u(k-p) + \beta T_{in}(k+i-n). \tag{F.2}$$

Insert (F.2) in the definition of the cost J_k given by (5.14) to get

$$J_k = \sum_{i=1}^{T} \left[\Delta T^*_{\text{out}}(k+i) - \alpha i R(k)u(k) - \alpha\bar{\varphi}(k, i)\right]^2, \tag{F.3}$$

where

$$\Delta T^*_{\text{out}}(k) \overset{\triangle}{=} T^*_{\text{out}}(k) - \beta T_{in}(k-n) \tag{F.4}$$

J. M. Lemos et al., *Adaptive Control of Solar Energy Collector Systems*, Advances in Industrial Control, DOI: 10.1007/978-3-319-06853-4,

and

$$\bar{\varphi}(k, i) \triangleq \sum_{p=1}^{n-i} R(k-p)u(k-p). \tag{F.5}$$

The derivative of the cost with respect to the manipulated variable is

$$\frac{1}{2}\frac{\partial J_k}{\partial u(k)} = -\alpha R(k) \sum_{i=1}^{T} i \left[\Delta T^*_{\text{out}}(k+i) - \alpha\bar{\varphi}(k, i)\right] + \alpha^2 R^2(k) u(k) \sum_{i=1}^{T} i^2. \tag{F.6}$$

Equating to zero the derivative (F.5) and solving with respect to $u(k)$ yields (5.16).

□

F.2 Proof of the WARTIC-State Control Law

Hereafter we prove Eq. (5.42) that yields a closed-form expression for the WARTIC-state control law.

From the assumptions (5.40) and (5.41) that both the manipulated variable and the reference are constant over the control horizon, the predictive model for the output (5.39) becomes

$$y(k+i) = CA^i c(k) + \left\{ C \sum_{j=0}^{i-1} A^{i-j-1} B \right\} u(k). \tag{F.7}$$

The matrices A, B and C are defined in (5.27). The cost function (5.33) can then be expanded as

$$\begin{aligned} J_k = &\left(r(k) - CA^i x(k) - \eta_{i-1} u(k)\right)^2 + \cdots \\ &+ \left(r(k) - CA^T x(k) - \eta_{i-1} u(k)\right)^2 + \rho T u^2(k), \end{aligned} \tag{F.8}$$

where

$$\eta_i = \sum_{j=0}^{i} CA^{i-j} B = \sum_{j=0}^{i} CA^j B = 1 + \beta + \cdots + \beta^i, \tag{F.9}$$

with $beta$ the parameter introduced in (5.13). Observing that

$$CA^i x(k) = \beta^i x_{n-i}(k), \tag{F.10}$$

gives

$$
\begin{aligned}
J_k = &\{\eta_0^2 + \ldots + \eta_{T-1}^2 + \rho T\}u(k)^2 \\
&+ 2\{\eta_0 x_n(k) + \eta_1 \beta^1 x_{n-1}(k) + \ldots + \eta_{T-1}\beta^{T-1} x_{n-T+1}(k)\}u(k) \\
&2r(k)\{\eta_0 + \eta_1 + \ldots + \eta_{T-1}\}u(k) + \text{terms independent of } u(k),
\end{aligned} \tag{F.11}
$$

and the $\min_{u(k)} J_k$ is obtained by solving the equation

$$
\begin{aligned}
\frac{1}{2}\frac{\partial J_k}{\partial u(k)} = &\left\{ \left\{ \sum_{i=0}^{T-1} \eta_i^2 \right\} + \rho T \right\} u(k) \\
&+ \sum_{i=0}^{T-1} \eta_i \beta^i x_{n-j}(k) + \left\{ \sum_{i=0}^{T-1} \eta_i \right\} r(k) = 0.
\end{aligned} \tag{F.12}
$$

Incorporating the constant T in ρ yields (5.42). □

Appendix G
Characterization of the Eigenvalues of the Linearized Tracking Dynamics

Proof of Proposition 6.1

The tracking dynamics is described by the following system of differential equations

$$
\begin{aligned}
\dot{x}_n &= 0 \\
\dot{x}_{n-1} &= \alpha R(1 - \tfrac{x_{n-1}-x_{n-2}}{x_n-x_{n-1}}) \\
\dot{x}_{n-2} &= \alpha R(1 - \tfrac{x_{n-2}-x_{n-3}}{x_n-x_{n-1}}) \\
&\cdots\cdots\cdots\cdots\cdots\cdots \\
\dot{x}_1 &= \alpha R(1 - \tfrac{x_1}{x_n-x_{n-1}}).
\end{aligned}
\tag{G.1}
$$

The first equation is already linear and provides an eigenvalue at the origin. Linearizing the other $n-1$ equations around the equilibrium point defined by

$$
x_i = \frac{r}{n} i \qquad r = 1
$$

yields the $(n-1) \times (n-1)$ Jacobian matrix:

$$
J = \begin{bmatrix} -2n & n & 0 & 0 & \ldots & 0 \\ -n & -n & n & 0 & \ldots & 0 \\ -n & 0 & -n & n & \ldots & 0 \\ \vdots & \vdots & \vdots & \vdots & \vdots & \vdots \\ -n & 0 & \ldots & \ldots & 0 & n \end{bmatrix}. \tag{G.2}
$$

In order to compute the eigenvalues of this Jacobian matrix, start by observing that it can be written as

$$
J = n(-I + A) \tag{G.3}
$$

where I is the identity of order $n-1$ and $A[(n-1) \times (n-1)]$ is the matrix

J. M. Lemos et al., *Adaptive Control of Solar Energy Collector Systems*, Advances in Industrial Control, DOI: 10.1007/978-3-319-06853-4,

$$A = \begin{bmatrix} -1\,1\,0\dots0 \\ -1\,0\,1\dots0 \\ \vdots\;\vdots\;\vdots\;\vdots\;\vdots \\ -1\,0\,0\dots0 \end{bmatrix}. \tag{G.4}$$

The characteristic polynomial of J is obtained from

$$\begin{aligned} \det(sI - J\frac{1}{n}) &= \det(sI + I - A) \\ &= \det((s+1)I - A) \\ &= \det(\sigma I - A), \end{aligned}$$

where the change of variable

$$\sigma := s + 1 \tag{G.5}$$

has been made.

Since A is a matrix in companion form, the coefficients of its characteristic polynomial (in σ) are the symmetric of the entries of the first column of A. Therefore:

$$\det(\sigma I - A) = \sigma^{n-1} + \sigma^{n-2} + \cdots + 1$$

Since (sum of the first $n+1$ terms of a geometric series):

$$\sum_{i=0}^{N} a^i = \frac{1 - a^{N-1}}{1 - a}, \quad \forall a \neq 1 \tag{G.6}$$

it is concluded that

$$\det(\sigma I - A) = \frac{\sigma^n - 1}{\sigma - 1} \quad \sigma \neq 1 \tag{G.7}$$

Hence, the eigenvalues of A satisfy

$$\sigma^n - 1 = 0 \;\; and \;\; \sigma \neq 1 \tag{G.8}$$

Recalling the change of variable (G.5), it is concluded that the eigenvalues of $\frac{1}{n}J$ are located over a circumference of radius 1, centered at -1. Furthermore, since

$$\lambda(J) = n\lambda\left(\frac{1}{n}J\right),$$

where $\lambda(M)$ denotes the eigenvalues of a generic matrix M, it follows that the eigenvalues of J are located over a circumference of radius n centered at $-n$.

In order to get the eigenvalues of the tracking dynamics, to the $n-1$ eigenvalues of J one must add another eigenvalue, located at the origin, associated with x_n.

In conclusion, there are n eigenvalues of the tracking dynamics, that divide the circumference in equal parts. □

Appendix H
Stability of a Time Varying System

Consider the linear time varying equation

$$\dot{e}(t) = \bar{A}u_*(t)e(t), \tag{H.1}$$

with

$$0 < u_{\min} \leq u_*(t) \leq u_{\max} \quad \forall t \geq 0 \tag{H.2}$$

and such that $\bar{A} = -A/L$ is a stability matrix with $\bar{\lambda}$ a negative number greater than the largest real part of its eigenvalues. Use the Gronwall-Bellman inequality Rugh (1996) to conclude that

$$\|e(t)\| \leq \|e(0)\| \exp\{\int_0^t \|\bar{A}u_*(\tau)\| d\tau\}. \tag{H.3}$$

Considering (H.2), this implies

$$\|e(t)\| \leq \|e(0)\| e^{\bar{\lambda} u_{\min} t}. \tag{H.4}$$

Since $\bar{\lambda} < 0$, this establishes asymptotic stability of (H.1). □

References

Goodwin GC, Sin KS (1984) Adaptive filtering prediction and control. Prentice-Hall, New York

Ibragimov NH (1999) Elementary Lie group analysis and ordinary differential equations. Wiley, New York

Mosca E, Zappa G, Lemos JM (1989) Robustness of multipredictor adaptive regulators: MUSMAR. Automatica 25(4):521–529

Rugh WJ (1996) Linear system theory, 2nd edn. Prentice- Hall, New York

J. M. Lemos et al., *Adaptive Control of Solar Energy Collector Systems*, Advances in Industrial Control, DOI: 10.1007/978-3-319-06853-4,

Index

J. M. Lemos et al., *Adaptive Control of Solar Energy Collector Systems*,
Advances in Industrial Control, DOI: 10.1007/978-3-319-06853-4,

MIX
Papier aus verantwortungsvollen Quellen
Paper from responsible sources
FSC® C105338

If you have any concerns about our products,
you can contact us on
ProductSafety@springernature.com

In case Publisher is established outside the EU,
the EU authorized representative is:
Springer Nature Customer Service Center GmbH
Europaplatz 3, 69115 Heidelberg, Germany

Printed by Libri Plureos GmbH
in Hamburg, Germany